Neurotransmitter Systems and their Clinical Disorders

Neurotransmitter Systems and their Clinical Disorders

Edited by

N.J. LEGG

Department of Neurology
Royal Postgraduate Medical School
Hammersmith Hospital
London, UK

1978

ACADEMIC PRESS

London · New York · San Francisco

A Subsidiary of Harcourt Brace Jovanovich, Publishers

ACADEMIC PRESS INC. (LONDON) LTD.
24/28 Oval Road,
London NW1

United States Edition published by
ACADEMIC PRESS INC.
111 Fifth Avenue
New York, New York 10003

Library of Congress Catalog Card Number: 78–72553
ISBN: 0–12–443050–3

Printed in Great Britain by
Whitstable Litho Ltd., Whitstable, Kent.

LIST OF AUTHORS

Sir Roger Bannister: The National Hospital, Queen Square, London WC1, U.K.

E.D. Bird: Neurochemical Pharmacology Unit, Department of Pharmacology, The Medical School, Hills Road, Cambridge, U.K.

D.M. Bowen: Department of Neurochemistry, Institute of Neurology, Queen Square, London, WC1, U.K.

D.W. Chadwick: University Department of Neurology, King's College Hospital and the Institute of Psychiatry, Denmark Hill, London SE5, U.K.

B. Costall: Postgraduate School of Studies in Pharmacology, University of Bradford, Richmond Road, Bradford, West Yorkshire, U.K.

T.J. Crow: Division of Psychiatry, Clinical Research Centre, Northwick Park Hospital, Harrow, Middlesex, U.K.

E.G. Gray: National Institute for Medical Research, The Ridgeway, Mill Hill, London NW7, U.K.

M. Hallett: University Department of Neurology, King's College Hospital and the Institute of Psychiatry, Denmark Hill, London SE5, U.K.

P. Jenner: University Department of Neurology, King's College Hospital and the Institute of Psychiatry, Denmark Hill, London SE5, U.K.

E.C. Johnstone: Division of Psychiatry, Clinical Research Centre, Northwick Park Hospital, Harrow, Middlesex, U.K.

S.Z. Langer: Laboratoires d'Etudes et de Recherches Scientifiques, 58 Rue de la Glaciere, Paris 75013, France.

C.A. Marsden: Department of Physiology and Pharmacology, The Medical School, Clifton Boulevard, Nottingham, U.K.

C.D. Marsden: University Department of Neurology, King's College Hospital and the Institute of Psychiatry, Denmark Hill, London SE5, U.K.

B. Meldrum: University Department of Neurology, King's College Hospital and the Institute of Psychiatry, Denmark Hill, London SE5, U.K.

R. Naylor: Postgraduate School of Studies in Pharmacology, University of Bradford, Richmond Road, Bradford, West Yorkshire, U.K.

C. Pallis: Royal Postgraduate Medical School, Du Cane Road, London W12, U.K.

J.D. Parkes: University Department of Neurology, King's College

Hospital and the Institute of Psychiatry, Denmark Hill, London SE5, U.K.

A.J. Pinching: Royal Postgraduate Medical School, Du Cane Road, London W12, U.K.

P.A. Price: University Department of Neurology, King's College, Hospital and the Institute of Psychiatry, Denmark Hill, London SE5, U.K.

C. Pycock: Department of Pharmacology, The Medical School, University of Bristol, Bristol BS8 1TD, U.K.

P.S. Sever: Medical Unit, St. Mary's Hospital, Paddington, London W2, U.K.

D.A. Shaw: Biochemical Psychiatry Laboratory, Department of Psychological Medicine, Whitchurch Hospital, Cardiff, Wales, U.K.

C.R. Snell: Laboratory for Peptide Chemistry, National Institute for Medical Research, Mill Hill, London NW7, U.K.

EDITOR'S NOTE

This volume is based on a meeting held by the Department of Neurology at the Royal Postgraduate Medical School in 1977. Neurotransmitter systems are one of the most important topics of study in neurology at present, and some of their disorders, such as the defect of dopaminergic neurones in Parkinson's disease, are amenable to effective treatment based on an understanding of the underlying pharmacopathology. The subjects covered here are necessarily selected. They include reviews of synaptic ultrastructure and of adrenergic transmission, defects of neuromuscular transmission and of the autonomic nervous system, abnormal movements, epilepsy, dementia, and behavioural disorders. These topics cannot be neatly classified according to the transmitter affected, for in only a few instances is this a simple matter, and the range of subjects reflects the range of precision with which the nature of the disorders is at present understood. Thus it spans myasthenia gravis at one extreme, where the transmitter is well known and we have progressed towards an understanding of the cause; through the abnormal movement disorders, where the once simple dopaminergic defect of Parkinson's disease has now to be viewed in a very complex context; to the more hypothetical accounts of behavioural abnormalities and speculation about the roles of the more recently discovered peptide transmitters.

I would like to thank all those concerned in the planning of the original meeting, in particular my co-organiser Dr. C. Pallis, and Dr. J. Reid and Prof. C.D. Marsden for their advice; all the contributors; and the publishers for their patience.

N.J. Legg

CONTENTS

SYNAPTIC ULTRASTRUCTURE

E.G. GRAY

*Department of Anatomy, University College London
Gower Street, London, U.K. ★*

Introduction

In this chapter I shall keep away from my own specialised field, which is to do with the way in which vesicles which we know to contain neurotransmitter actually arrive at the presynaptic membrane. Instead, I shall try to put over some of the general concepts which people are grappling with nowadays.

First of all I will say something about the fine structure of chemical synapses, and then will go on to discuss the relations which we are beginning to discover between the electron microscopic appearance of a synapse and the transmitter which is involved. Much of this is new work, and some of it is controversial, so I will have to abbreviate a great deal and be somewhat dogmatic.

Fine Structure of Chemical Synapses

By electron microscope (EM) techniques we can now distinguish various groups of chemical synapses: the first are cholinergic; the second are the amine group, in which are included noradrenergic endings and those which employ dopamine and 5-hydroxytryptamine; and the third group are the amino acids, including glutamate, which many people suspect is an excitatory transmitter, and the two well known inhibitory amino acids, gamma amino butyric acid (GABA) and glycine. Acetylcholine can of course be either excitatory or inhibitory, at different sites, and the same is true for the aminergic synapses. GABA is in fact known to be depolarising under certain circumstances, but we have to keep things simple for the moment, so I am not going to discuss the implications of presynaptic inhibition and related phenomena. Let us then consider the ultrastructural appearances of different chemical synapses, and the ways in which an understanding of these is beginning to point the way to a neuro-

★Present address: National Institute for Medical Research, Mill Hill, London, U.K.

chemical mapping of the central and peripheral nervous systems. Such mapping has, in fact, been very fully carried out in the cerebellar cortex, but the much more complex cerebral cortex is a different matter, and we are just beginning to lay down the rules by which we may infer various forms of synaptic interaction.

Figure 1 shows an EM picture of a large brain stem neuron, with a nucleus (n) and copious mitochondria (m). The axon is very tortuous, and this is true of both myelinated and unmyelinated fibres: this makes it very difficult at ultrastructural level to be certain in any one section of more than a few synaptic connections. Synapses

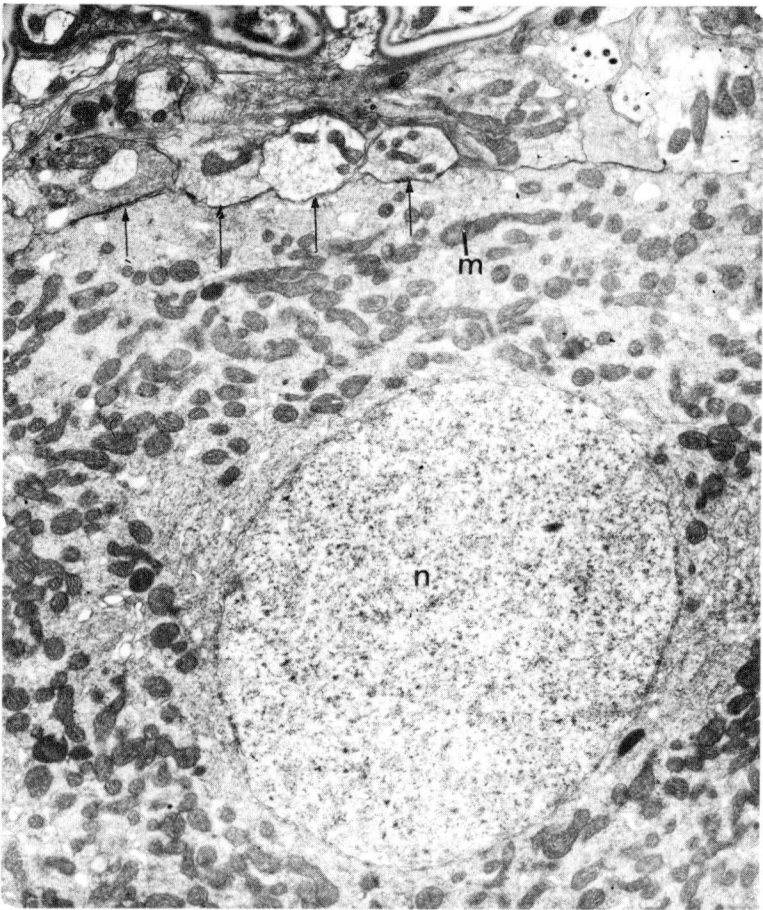

Fig. 1 Large brainstem neuron with boutons (arrows). m = mitochondrion, n = nucleus. x 3,500.

(arrows) are of course found not only on the cell body but also on the dendrites, which are not apparent in the figure.

Figure 2 illustrates a dendrite, with mitochondrion (m) and this is one of those rare examples where one can follow the axon (a) in, and see it bulging out into its terminal varicosity and making contact with the dendrite. Now for some of the details which can be made out in this picture. These include the mitochondria: mitochondria are of

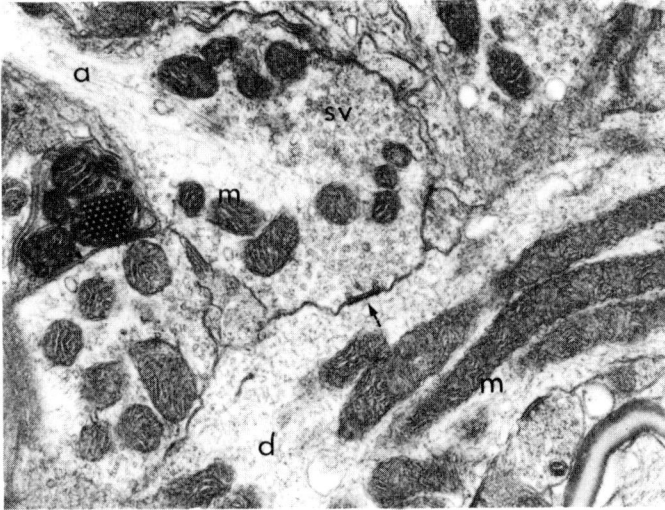

Fig. 2 Bouton synapsing on a dendrite. a = axon, m = mitochondria, sv = synaptic vesicles, d = dendrite. x 10,000.

course ubiquitous, but they are often seen in the presynaptic knob. The more important and typical feature, however, is the presence of synaptic vesicles (sv), of more or less constant size, with a diameter of about 500 Angstroms (see also Fig. 3). Because of their constant size, and for a number of other reasons, they are thought to contain the neurotransmitter, whichever it may be, in quantal amounts. Where the presynaptic membrane abuts on to the dendrite, there is usually a thickened region, with an aggregation of vesicles near it, and this is becoming known as the 'active zone' (arrow). It is assumed that when the impulse arrives at the terminal it is from this site that the neurotransmitter in the vesicle is released across the synapse.

Figure 3 shows two synapses. The right one does not show the complete axon coming in (y), but a small portion of it is apparent, and it is unmyelinated. When the nerve impulse travels down the axon it does so over its surface, and when it reaches the region of the synaptic cleft it induces calcium transport across the membrane,

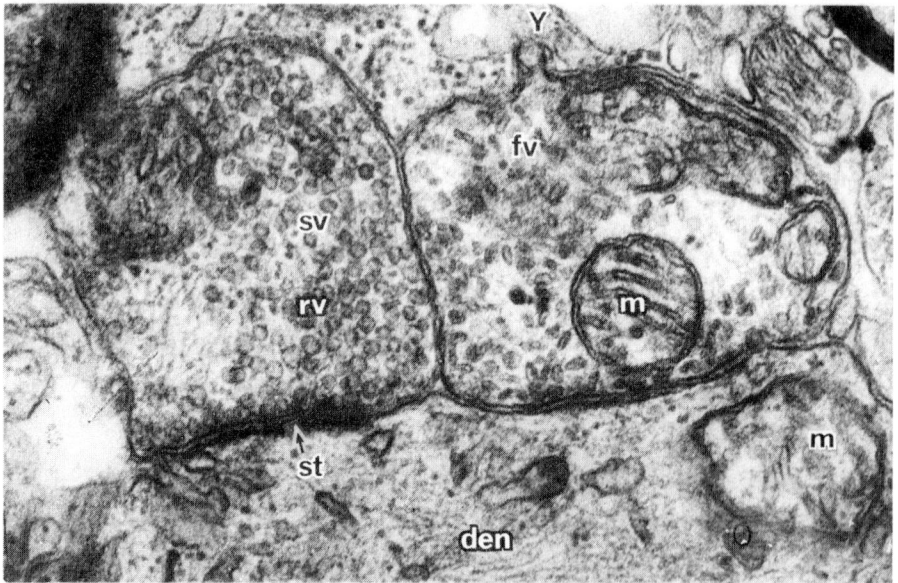

Fig. 3 Two boutons synapsing on a dendrite (den) in spinal cord. The left one has round vesicles (rv) and is probably an excitatory synapse, and the right one has flat vesicles (fv) and is probably inhibitory. The left one has pronounced type 1 thickening (st). x 37,000.

into the terminal, and this is characteristic of all chemical synaptic mechanisms, and also of neurosecretory and exocrine secretory cells. The action of the calcium is unknown but it seems vital to the release of transmitter, which, as a result of opening of the vesicles, passes across the synaptic cleft to receptors on the post-synaptic membrane. These have not yet been identified by electron microscopy; it has however, proved possible to isolate them with the help of alpha-bungaro-toxin, but this is a biochemical method and so far has not been correlated with ultrastructure. In a cholinergic synapse, the acetylcholine is inactivated by the action of cholinesterase, which breaks it down into acetic acid and choline, the latter being reutilised in the presynaptic bulb. In the aminergic endings the transmitter is not broken down, but there is a remarkably efficient re-uptake mechanism, well demonstrated for noradrenalin, which is a very efficient way of conserving neurotransmitter.

The synaptic delay, which is of the order of a millisecond, is somehow related to the entry of calcium and the release of neuro-transmitter: the actual diffusion of neurotransmitter across the synaptic cleft, which is only about 200 Angstroms wide, can be measured in microseconds, and does not therefore make a substantial contribution to the delay. After the transmitter has combined with the receptor, it brings about either a depolarisation or a hyper-polarisation, depending on the transmitter and the nature of the synapse.

Relation between Structure and Fuction

Now for some of the problems which neuroscientists are wrestling with at present. First of all there is strong evidence that the vesicles are real *in vivo*, although even this is being questioned. Naturally the whole theory of quantal release would fall down if the vesicles were not a structural reality in living tissue (e.g. formed artifactually from a labile tubular system). One quantum is, of course, supposed to correspond with one vesicle, and the standard size of the vesicles is in accord with this. Furthermore the vesicles form aggregations on the presynaptic membrane, in the ideal strategic position for the release of transmitter, and of course there is much other evidence, for example from vesicle fractions prepared by ultracentrifugation, that they do in fact contain the transmitter. One problem, however, is what the vesicles do when they have released the transmitter: do they fuse with the membrane, or do they return inside the terminal to be recharged with transmitter? If they fuse with the membrane this would produce a large increase in membrane area, and one hypothesis suggests that there is a recycling mechanism, such that each time a vesicle fuses with the membrane another one is pinched off from an adjacent part of the membrane to form a new vesicle. This is a hypothetical and controversial mechanism at present and would of course involve a very high turnover of membrane, which is one of the reasons against the idea. In any case the vesicle has at some point to be recharged with transmitter, which may be taken up across the surface membrance as the complete transmitter in the case of noradrenalin, or as choline in the case of acetylcholine. A problem which particularly interests me, as I mentioned earlier, is how the vesicles make their way to the active zone on the presynaptic membrane; do they come down from the cell body, or are they formed locally within the terminal, and in any case how do they line up and cluster round the active zone?

For the present let us assume that the vesicles do indeed contain quanta of transmitter, and proceed from there.

One can examine literally thousands of synapses with the EM before seeing what looks like a vesicle discharging into the synaptic cleft. The evidence for the opening of vesicles cannot of course depend on a few isolated micrographs, and in fact important evidence comes from the technique of freeze fracture. I will not go into the technique of this in detail, but it is possible nowadays to fracture the presynaptic membrane in such a way that you can look at its inner surface. This is done at extremely low temperature and under a high vacuum, and a metal spraying technique will then produce a high quality replica of the surface, which can be examined with the EM.

Such replicas of brain synapses show the presence in the membrane of pores, which were at first known as synaptopores, but which have more recently been called vesicle sites or vesicle attachment sites. In this picture one cannot of course see inside the presynaptic bag, for this method only permits one to look directly at the shadowed inside of the presynaptic membrane. From this type of observation it could be inferred that the pores were in fact lying opposite where the vesicles were situated on the presynaptic membrane. This has led many people to believe that the pores are specifically related to the sites of vesicle release. However, there are a number of problems still to be sorted out: a detailed counting, for example, has shown that there are far too many synaptopores to account for the number of vesicles which would have been discharging at the time of fixation. Another difficulty is that the synaptopores cannot be seen if fixation is carried out rapidly under certain conditions. None of this evidence, in any case, indicates how the vesicles relate to the membrane. Some believe that the vesicle comes down and discharges its contents through the membrane pore, and then moves back into the presynaptic bag to be recharged with transmitter. However, others think that the vesicle membrane fuses with the presynaptic membrane, which then leads to the recycling and high membrane turnover which we mentioned above.

Tying down a neurotransmitter to a vesicle at a given synapse is, of course, the thing we would like to do, and the nicest way would be to have a specific stain for each neurotransmitter and then examine sections under EM to discover whether a particular vesicle has taken up a particular stain. Unfortunately this technique is still only a remote possibility. The first line of attack (Fig. 4) was to use homogenised whole brain and then fractionate it (by ultracentri-fugation) into various size particles. By this method a very satisfactory preparation of nerve endings, which are very tough (sometimes with post-synaptic membrane attached to them) was made. Then by further hypertonic treatment and high speed ultracentrifugation an almost pure preparation of synaptic vesicles could be obtained and analysis of this fraction showed the presence of the various neurotransmitters, depending on which part of the brain or spinal cord had been selected

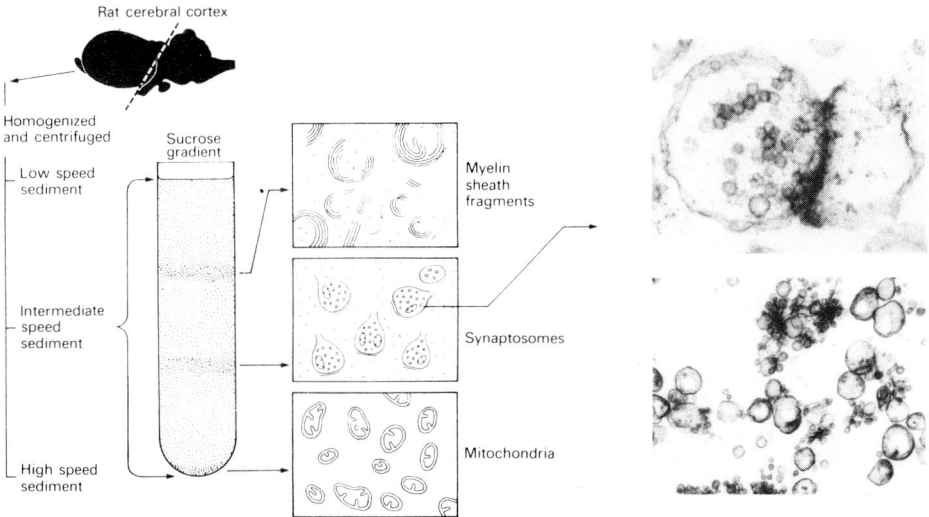

Fig. 4 Method for preparation of synaptic vesicles by brain fractionation.

for processing. This demonstration of transmitters within the vesicle fraction was the first concrete evidence that the vesicles themselves did indeed contain neurotransmitter, but the big disadvantage of the method is that one has to use large pieces of nervous tissue, such as the whole of the forebrain, so that the collection of presynaptic bags is very heterogeneous and it is impossible to learn anything about which synapse contains a specific transmitter.

Localisation of Transmitters

In order to refine our analysis we need to look at sections of brain rather than homogenised regions, and the idea is to see how far mapping can be carried out using a combination of EM and histo-chemical methods.

Cholinergic Synapses

Unfortunately acetylcholine synapses are the most difficult to recognise; in fact we are as yet unable to identify a cholinergic synapse in the central nervous system, by light or electron microscopy, and be absolutely certain that it is cholinergic. The only place where we can be certain of having a cholinergic nerve ending is at the vertebrate motore end-plate, and this has been used for experimenting with a variety of staining techniques. In the spinal cord, for example, it is still impossible to say which are the cholinergic synapses from the recurrent collaterals of the ventral horn cells on to Renshaw cells, and although there must be many cholinergic synapses within the cerebral cortex we have not been able to identify them either. This difficulty with the best known and most studied neurotransmitter is thus a major obstacle to mapping the CNS at present. Some regions of the brain are rich in cholinergic endings, and Akert and his co-workers used a modification of an old Champy technique, known as the zinc-iodide-osmium method, which they applied to this region. With conventional processing the vesicles look empty, but by this technique they were found to contain electron dense material (Fig. 5). Since this was observed specifically in a cholinergic-rich area it was initially

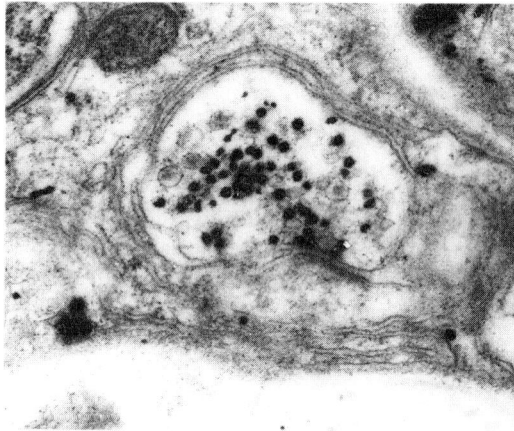

Fig. 5 Autonomic synapse. The vesicles show a black impregnation of zinc-iodide-osmium. x 20,000.

supposed that this technique might prove to be a specific stain for cholinergic nerve endings, so the method was tried out on the motor end-plate with similar results. Various other workers, equally excited by the possibility, used the technique for staining other brain areas known to be rich in other transmitters such as noradrenalin, GABA or glycine, but unfortunately they found that these synaptic vesicles also became impregnated by the zinc-iodide-osmium. Thus it became apparent that this was not a specific test for cholinergic endings, but since it reacts with something inside the synaptic vesicles, this may be of considerable interest in itself.

A less direct approach to the location of acetylcholine is to look for the presence of cholinesterase, and this was pioneered by the elegant work of Shute and Lewis. It is possible to localise cholinesterase in the synaptic cleft with the EM. This involves a modification of the method previously used for light microscopy. It is worth noting that cholinesterase is not only found in the synaptic cleft, but appears to be present all around the pre-synaptic ending, so we do not know where it is secreted from. This, then, is one indirect method of localising cholinergic synapses, but unfortunately workers have now been trying this technique on the spinal cord synapses, which are known not to be cholinergic but to employ GABA or glycine, and these also sometimes show positive reactions for cholinesterase. This means that the method has to be treated with caution, but there is reasonable evidence that when the tracts that run to synapses also stain for cholinesterase then the synapse is most probably cholinergic.

We are thus still left with some uncertainty about the localisation of endings. However, some workers are progressing towards an EM method for localising choline acetylase (or choline acetyltransferase, as it is now called) which is a cytoplasmic enzyme specific for the formation of acetylcholine, and this may well prove a satisfactory method for identifying cholinergic synapses positively in the central nervous system.

Aminergic Synapses

These synapses fortunately are more amenable to study by the electron microscopist, because in many cases they have 'markers' in the vesicles. Figure 6 is a noradrenergic synapse from Auerbach's plexus, and it shows that many of the vesicles (arrows) contain dense particles, which are related to the presence of noradrenalin, or perhaps to one of the proteins which bind it — there is still some argument about this. Whatever the explanation, however, this does seem to be a marker for adrenergic synapses, and there have been various methods for checking this. One of these is the use of reserpine, which will over a few days deplete an animal's nerve endings of noradrenalin. The method therefore involves examining reserpinised animals' nerve endings by EM, and it is found that the intravesicular dense particles have disappeared. A quite separate method of identifying adrenergic synapses is by light microscopy, using the Falck fluorescence technique. The synaptic preparation is treated with aldehyde gas and then examined with ultraviolet fluorescence microscopy, and a greenish or yellow fluorescence is seen, which is not found if the animal has been pre-treated with reserpine.

A further technique involves the use of the so-called false neurotransmitters. You will recall that the amine transmitters are not broken down like acetylcholine, but are taken up again by a very

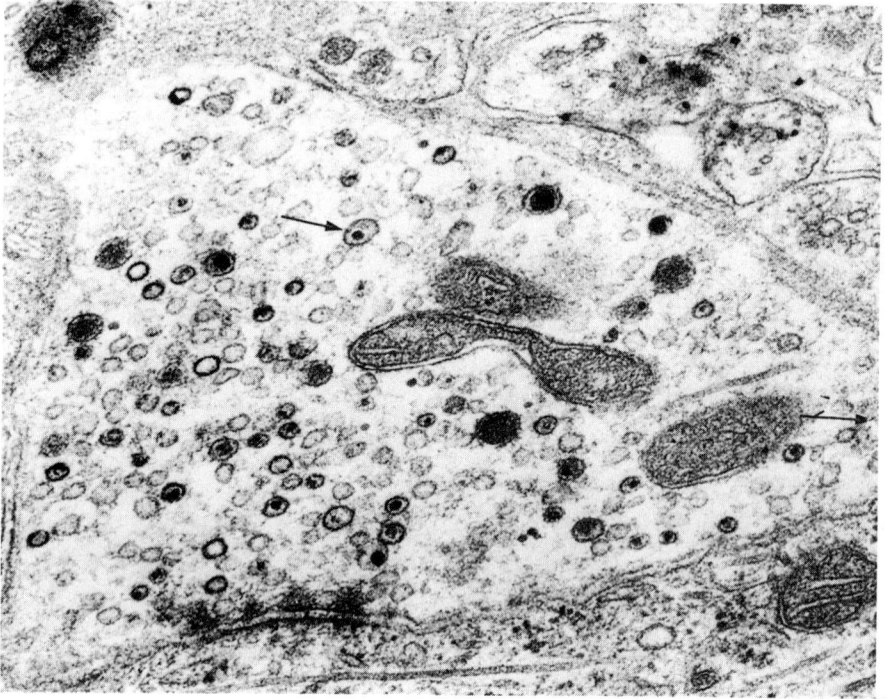

Fig. 6   Aminergic synapse. Vesicles contain a dense particle (arrowed). (By kind permission of Dr. G. Gabella). x 50,000.

efficient pump, and the false neurotransmitters are substances which are taken up in the same way and incorporated into vesicles, and can be released on stimulation. Five-hydroxydopamine and six-hydroxydopamine are both taken up in this way, and the technique used is to soak the suspected noradrenergic synapse in one of these substances, and then to examine the preparation under the electron microscope, when a high percentage of the vesicles will be found to contain dense granules. This is a particularly useful technique for certain smooth muscle preparations and other tissues in which conventional processing is not satisfactory. Six-hydroxydopamine is, in fact, a cellular poison because its metabolism involves the formation of hydrogen peroxide. Thus the use of this 'false transmitter' results in an *in vitro* degeneration technique, producing not only granules in the vesicles but ultimately death of the nerve terminal. This is therefore a very useful experimental method for sorting out different sorts of nerve endings.

Amino Acid Transmitters

In 1957 I began work on the cerebral cortex, using a stain which had not before been used on the CNS, and a new epoxy resin embedding technique for thin sectioning for electron microscopy. With these techniques, I described two sorts of synapse in the cerebral cortex, designated types 1 and 2: type 1 had an intensely staining region of proteinaceous material on the post-synaptic side of the cleft (Fig. 3, st) and type 2 (Fig. 3, right synapse) had very little such material. There were other differences: in type 1 synapses the post-synaptic thickening extended over almost the entire area of synaptic contact, whereas in type 2 the thickening was present in only small patches. There were also other differences that I will not go into here. Soon afterwards Eccles, Blackstadt and Anderson, who were working on the hippocampal cortex using single cell electrical recording from the large pyramidal cells, found that inhibitory synapses were situated on the cell bodies and excitatory synapses on the dendrites. When

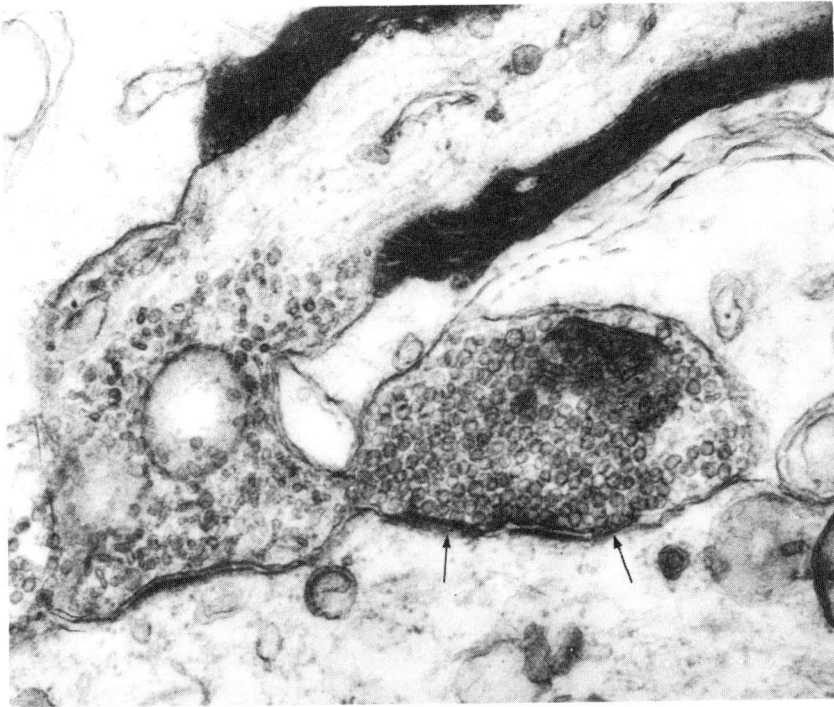

Fig. 7 Two boutons on a dendrite in fish spinal cord. The right one is an electrical synapse showing membrane 'fusion' (gap junctions) in two places (arrows). x 32,000.

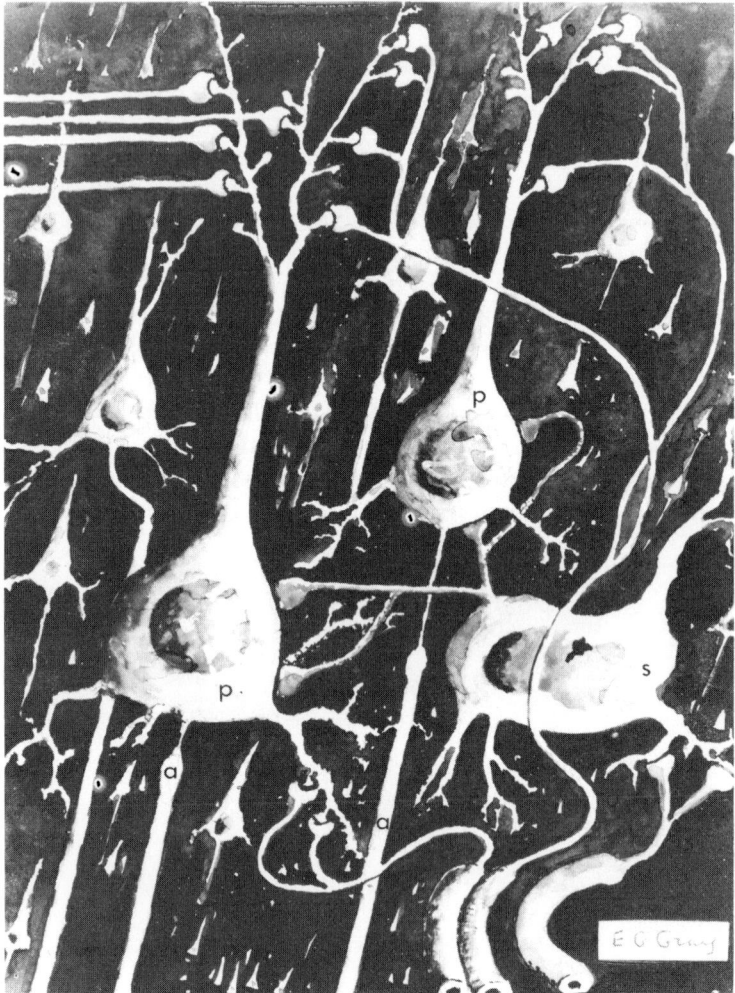

Fig. 8 Photograph (taken from a water colour by the author) of a diagram of a block of cerebral cortex.

they examined their material with the EM they found that all the synapses on the cell bodies were of type 2 and those on the dendrites were type 1. In 1963 Eccles therefore suggested that my type 1 were excitatory synapses and the type 2 were inhibitory ones. This was found to hold good for the cerebellum, where the physiology had also been well worked out, but it did not fit in well with observations on the spinal cord. The next step came in 1965, when Uchizono, using a special aldehyde technique, showed that excitatory synapses, with

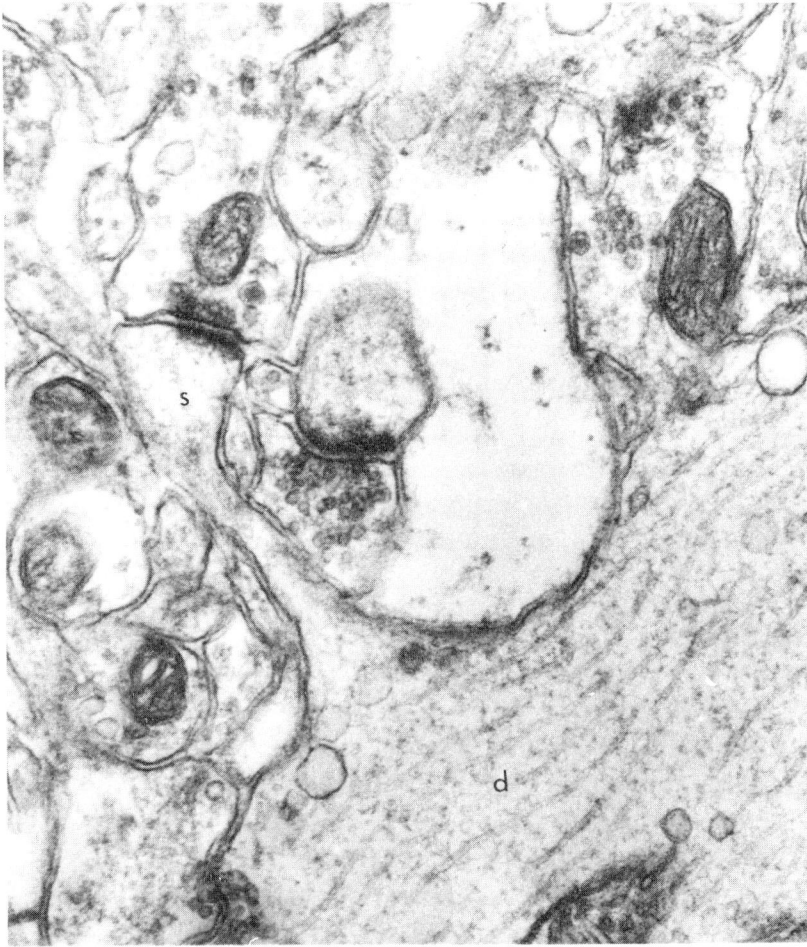

Fig. 9 Synapse on dendritic spine (s) in cerebral cortex. x 27,000.

type 1 morphology, had round vesicles, and inhibitory synapses, with
type 2 morphology, had flat vesicles. This observation was held up
in many CNS regions studied with the EM. Why the vesicles stored in
inhibitory terminals are flat is unknown at present.

Thus there are now very hopeful prospects for mapping the
cortex, because we may reasonably suppose that if we see a type 1
synapse with round vesicles, this is excitatory. In the cerebral cortex
glutamate may be the transmitter involved, whereas the type 2
synapses with flat vesicles, either in the cord or in the cortex,
probably utilise GABA or glycine. Matus and Denison have in fact

been able to show that the endings with flat vesicles can take up radiolabelled glycine, whereas the round vesicle endings will not.

Electrical Synapses

Electrical synapses (Fig. 7) are common in the CNS of lower vertebrates (fishes particularly) but rare in the mammalian CNS. These electrical synapses have fused synaptic membranes and also contain vesicles, and this was for a long time something of a puzzle, since there was evidence only for electrical transmission across these synapses and the presence of vesicles presented an anomaly. However, more recent work has shown that there is also a chemical excitatory transmission at these electrical synapses, so the presence of vesicles can now be accounted for.

Synaptic Organisation in the Cortex

Colonnier used Uchizono's method on the cerebral cortex and showed that there were two populations of nerve endings, round-vesicled and flat-vesicled. As a result we can now begin to see a pattern in the cerebral cortical connections. There are, as has been known for a long time, two main cell types, and these are illustrated in Fig. 8: pyramidal cells (p) and stellate cells (s). The pyramidal cells are characterised by dendritic spines, as shown in Fig. 9 (s). Synapses on these spines are of type 1 with round vesicles (Fig. 10, left), and are presumably excitatory; those on the soma are of type 2 with flat

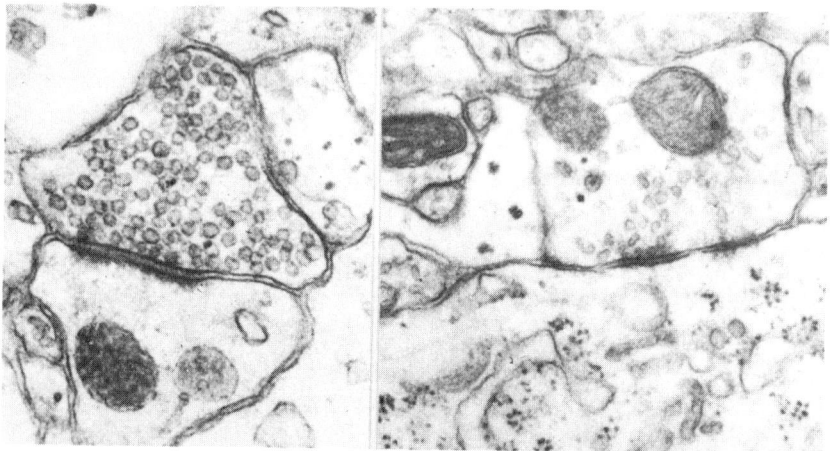

Fig. 10   Synapses in cerebral cortex. Left — synapse on a dendritic spine with round vesicles and type 1 thickening. Right — synapse on the soma with flat vesicles and type 2 thickenings. (By kind permission of Prof. M. Colonnier). x 35,000.

vesicles (Fig. 10, right), and are presumably inhibitory. Pyramidal cells themselves are probably always excitatory, and they often send their axons outside the local cortex (Fig. 8, a). According to Dale's principle their recurrent collaterals should also be excitatory, with type 1 thickenings and round vesicles. The small stellate cells (Fig. 8, s), by contrast, and also the large stellate cells, are all inhibitory; their axons, which often end on the soma of a pyramidal cell, therefore have type 2 thickenings and flat vesicles (Fig. 10, right). All incoming fibres into a block of cortex may well be excitatory. Many will end on pyramidal cell dendritic spines, but whether they end here or on stellate cells they will have type 1 thickenings and round vesicles. It is on this sort of basis and from data such as these that we are beginning to understand the ultrastructure of the cerebral cortex.

Bibliography

1976. The Synapse. Cold Spring Harbor Symp. Quant. Biol., Vol. 40. pp. 694.

Corner, M.A. and Swaab, D.F. (1976). Perspectives in brain research. *Prog. Brain Res.*, **45**, 1–488.

Palay, S.L. and Chan-Palay, V.L. (1974). Cerebellar Cortex. pp. 348, Springer, Berlin.

Papas, G.D. and Purpura, D.P. (1972). Structure and function of synapses. pp. 308. North Holland Publ. Co., Amsterdam.

Peters, A., Palay, S.L. and Webster, H. de F. (1976). The fine structure of the nervous system. pp. 406. W.B. Saunders, London.

DISEASES OF THE NEUROMUSCULAR JUNCTION:

Pathophysiological mechanisms in Myasthenia Gravis and the Eaton-Lambert Syndrome

A.J. PINCHING

Department of Medicine, Hammersmith Hospital, London, U.K.

Introduction

The neuromuscular junction has, from the investigator's point of view, the considerable advantage of being a discrete neurotransmitter entity with easily identifiable afferent and effector components. Its structural and functional properties have been intensively studied; myasthenia gravis and the Eaton-Lambert (or myasthenic) syndrome, which are both known to be due to disordered neuromuscular transmission, are well characterised clinically. It is perhaps surprising then that it has only been in the last few years that the precise nature of the functional disturbance in these diseases and their pathogenetic basis have been elucidated, especially in the case of myasthenia gravis (Drachman, 1978).

Clinical Features of Neuromuscular Diseases

Clinically, myasthenia gravis is characterised by weakness and fatiguability of skeletal muscle which may affect different muscles to varying degrees. Disease restricted to the ocular muscles is common but the more severe generalised form affects the limbs, especially the proximal muscles, and the vital bulbar and respiratory muscles. Two main groups of patients are affected: a younger group with a female preponderance often associated with thymic hyperplasia, and an older group with male preponderance frequently associated with thymomas. The electrophysiological hallmark of the disease is a decremental response to repetitive stimulation, the electrical counterpart of fatiguability.

The Eaton-Lambert or myasthenic syndrome on the other hand is a disorder affecting predominantly limb muscles and is generally associated with small cell carcinoma of the bronchus. Weakness is maximal at the onset of a movement and strength increases as a movement is continued or repeated. Electrophysiologically this is mirrored by post-tetanic potentiation. While both of these disorders are due to disordered function of the neuromuscular junction, they are so different in their clinical features and their pathogenesis that their nominal similarity is unfortunate.

Physiology of Normal Neuromuscular Transmission
The release of 'packages' of acetylcholine following electrical depolarisation of the nerve terminal is the familiar first event of chemical transmission from nerve to muscle. Acetylcholine molecules, after diffusing across the intersynaptic cleft, reach specific acetylcholine receptors on the postsynaptic membrane. These receptors are large glycoprotein molecules, which after reversible interaction with acetylcholine lead to alterations in ionic channels in the endplate membrane; this effect is mediated by the ionophore, which either forms part of the acetylcholine receptor molecule or is a closely adjacent molecule. Sufficient ion flux leads to depolarisation of the muscle membrane and fibre contraction. The released acetylcholine is either inactivated by acetylcholinesterase or diffuses away from the junction; in this way the chemical message dispatched by a given nerve impulse is brought to a full stop. The release of acetylcholine is usually rather generous in terms of the threshold required to achieve depolarisation; this provides a 'safety factor' for transmission.

Intracellular recording in the region of the motor endplate is used to study potential changes resulting from acetylcholine release. The co-ordinated release following a nerve impulse leads to an endplate potential which, if it reaches a given threshold level, will lead to a propagated potential in the fibre. The spontaneous release of quanta of acetylcholine without nerve stimulation leads to tiny depolarisations known as miniature endplate potentials (m.e.p.p.'s). Analysis of the resting membrane potential shows fluctuations (acetylcholine 'noise') which are considered to represent the opening of ion channels via the ionophore following the interaction of acetylcholine with the acetylcholine receptor.

Myasthenia Gravis
The Physiological Defect
The site of the physiological defect in myasthenia gravis — whether pre- or post-synaptic — has been the subject of controversy for some time; the tentative conclusion of Elmqvist *et al* (1964) that it was presynaptic was widely accepted. These workers showed that the

m.e.p.p.'s were reduced in amplitude, consistent with decreased size of acetylcholine quanta; they were unable to demonstrate alteration in postsynaptic sensitivity using carbachol and decamethonium. Much recent work, however, (summarised below) has indicated that the disorder is in fact postsynaptic. It is clear that the alteration in m.e.p.p. amplitude would be consistent with either a disorder of acetylcholine release or a defect in its postsynaptic action.

The presynaptic structures of the neuromuscular junction in patients with myasthenia gravis are seen to be normal on ultrastructural examination; in particular the synaptic vesicles are normal in number and size. Striking postsynaptic changes are seen, however, with widening of the intersynaptic cleft and simplification of the membrane infoldings (Engel and Santa, 1971).

Investigation of the acetylcholine content of the neuromuscular junction using gas chromatography and mass spectroscopy has shown that this is normal or even increased (Ito *et al*, 1976). There is no evidence of a false transmitter (an alternative suggestion by Elmqvist *et al*, 1964). Furthermore, acetylcholine release following chemical depolarisation is the same as in normals; the resting release of acetylcholine from muscle is the same as in controls (Ito *et al*, 1976). The reduction in m.e.p.p. frequency often seen in myasthenic muscles is explained by the fact that the small potentials are lost in the noise of the system (Ito *et al*, 1978b). The complementary study on postsynaptic sensitivity to iontophoretically applied acetylcholine has shown reduced sensitivity in myasthenic neuromuscular junctions (Albuquerque *et al*, 1976a). The acetylcholine noise of myasthenic and control endplates is similar, suggesting that the ionophore is not affected (Ito *et al*, 1978b).

Acetylcholine Receptors

The electrophysiological evidence is therefore very strongly in favour of an acetylcholine receptor defect. A more direct demonstration of this derives from studies using α-bungarotoxin; this is a snake toxin that binds specifically and irreversibly with the acetylcholine receptor and may be labelled with [125] I. Using autoradiography, Fambrough *et al* (1973) showed that the number of acetylcholine receptors thus labelled was substantially reduced in myasthenic neuromuscular junctions. This has been confirmed by several groups of workers, who have also shown a close correlation between the reduction in m.e.p.p. amplitude and the reduction in α-bungarotoxin binding (Engel *et al*, 1977a; Lindstrom and Lambert, 1978; Ito *et al*, 1978b).

There are two possible mechanisms that could account for a

postsynaptic defect resulting in decreased α-bungarotoxin binding: one is a decrease in the actual number of acetylcholine receptor molecules, and the other is that, although present, some of the receptors are blocked or prevented from functioning normally. The fact that the postsynaptic membrane is structurally simplified in myasthenia could be taken as favouring the former mechanism, and evidence has been adduced that some of the defect in neuromuscular transmission can be explained in terms of receptor loss (Lindstrom and Lambert, 1978).

Acetylcholine Receptor Antibodies

That acetylcholine receptor blocking could occur in myasthenia was first indicated by Almon *et al* (1974), who showed that the pre-incubation of a denervated rat muscle preparation with myasthenic serum (globulin) resulted in a decrease in α-bungarotoxin binding sites. This was later confirmed using a human muscle preparation (Bender *et al*, 1975). Subsequently a variety of techniques have demonstrated that the blocking factor is an autoantibody to acetylcholine receptor (Aharonov *et al*, 1975a; Appel *et al*, 1975; Lindstrom *et al*, 1976c; Mittag *et al*, 1976; Ito *et al*, 1978a). This antibody may be found in up to 87% of myasthenics by precipitating labelled acetylcholine receptor-antibody complexes with antihuman IgG (Lindstrom et al, 1976c). The level of antibody found in serum correlates only broadly with the severity of disease, the lowest levels being found in cases with ocular involvement only and the highest levels tending to occur in those with severe active disease (Lindstrom *et al*, 1976c; *Ito et al*, 1978a). Ultrastructural studies have provided suggestive evidence that gammaglobulin-like molecules are attached to receptor in myasthenia (Rash *et al*, 1976) and special staining methods have shown the presence of IgG and complement attached to acetylcholine receptor at myasthenic neuromuscular junctions (Engel *et al*, 1977b).

It has been relatively difficult until recently to find evidence that myasthenic serum could interfere with neuromuscular transmission in vitro, but by using suitable assay systems such an effect has now been seen (Anwyl *et al*, 1977; Bevan *et al*, 1977). While these various findings are strongly in favour of myasthenia gravis being due to a postsynaptic defect, in part mediated by fixation of auto-antibody to acetylcholine receptor, they are bound to fall short of proof unless it can be shown that the autoantibody does interfere with neuromuscular transmission in the myasthenic patient.

The Pathogenetic Significance of Acetylcholine Receptor Antibodies

The evidence that the acetylcholine receptor antibody is of pathogenetic significance has come from four separate but related lines of evidence. Transient neonatal myasthenia is a disorder closely resembling myasthenia that affects the offspring of some myasthenic mothers; the weakness is worst soon after birth and gradually improves over the next month or two (Namba *et al,* 1970). The implication that this disorder is due to transplacental transfer of some substance concerned in the disease of the mother is obvious, and the possibility that it is an immunoglobulin is suggested by the natural history and the duration (Simpson, 1960). Nature's passive transfer experiment has recently been shown to be mediated by the transplacental transfer of acetylcholine receptor antibody, whose disappearance from the infant's circulation closely parallels the clinical improvement (Keesey *et al,* 1977).

Bergstrom et al (1973, 1975) found that thoracic duct lymph drainage led to striking benefit in patients with myasthenia gravis; they showed that re-infusion of thoracic duct lymphocytes had no effect on the patients but that infusion of cell-free lymph led to a prompt deterioration in the clinical state. This demonstrates that the procedure, which might have been expected to have been effective through the removal of T lymphocytes, was in fact causing depletion of humoral factors. Subsequent work has confirmed that the active fraction is an immunoglobulin with anti-acetylcholine receptor activity (Lefvert and Bergstrom 1976).

A more convenient way of depleting humoral factors, including immunoglobulin, is by the use of plasma-exchange (Pinching, 1978). Initial studies demonstrated a clear-cut improvement in the clinical state of patients with acquired myasthenia gravis following plasma-exchange (Pinching *et al,* 1976) and these findings have been confirmed in larger series (Dau *et al,* 1977; Newsom Davis *et al,* 1978). Increased muscle power affecting all muscle groups starts two to three days after the first exchange and reaches its peak a similar interval after the last exchange. The improvement has been maintained for long periods, in some cases by immunosuppression.

Longitudinal studies on patients following plasma-exchange probably provide the strongest evidence that the alterations in the clinical state are due to changes in the level of acetylcholine receptor antibody. A prompt fall in the antibody titre occurs during the exchange period, along with comparable reductions in total immunoglobulin and, for example, fibrinogen and complement levels.

The rate of recovery of the various plasma components shows that whereas total IgG, fibrinogen and complement return to normal relatively rapidly, as would be expected from their known synthetic rates, the levels of specific antibody do not generally return until much later; they show a striking relationship with the clinical state, the measured indices of muscle strength mirroring closely the antibody levels (Dau et al, 1977; Newsom Davis et al, 1978). The duration of remission can be correlated with the rate of return of antibody; this in turn varies with the duration of the disease and the use of immunosuppressive agents. Patients with active antibody synthesis in whom plasma-exchange is performed without subsequent immunosuppression show early deterioration, sometimes becoming weaker than before exchange, associated with a rapid rise, and sometimes rebound, of antibody levels.

While the level of circulating acetylcholine receptor antibody in different patients correlates relatively poorly with the amount bound to the endplate (Lindstrom and Lambert, 1978) — as would be expected from the variable avidity of antibodies — in such longitudinal studies on single patients it is likely that the equilibrium between fixed and free antibody has been disturbed in a predictable way. It is not certain however whether the recovery following removal of circulating antibody is due to net loss of antibody fixed to the receptor or due to the synthesis of new receptor free from antibody. Observations showing that the rate of receptor degradation is increased in the presence of antibody (Appel et al, 1977; Kao and Drachman, 1977b) favour the latter explanation.

A further line of evidence favouring the role of specific antibody in the pathogenesis of myasthenia gravis has been the work of Toyka et al (1977), in which the passive transfer of myasthenia from man to mouse has been effected. The transfer was achieved with an IgG antibody and was dependent on complement C3 but not on the late complement components.

The postsynaptic defect in myasthenia gravis appears therefore to result from a combination of decreased amount (40%) of acetylcholine receptor (associated with structural changes at the endplate) with impaired function in the remaining receptors due to antibody fixation (Lindstrom and Lambert, 1978); this acetylcholine receptor antibody does not fix at the acetylcholine or α-bungarotoxin binding site of the receptor, but seems to interfere by an allosteric effect from a neighbouring antigenic site on the molecule (Almon and Appel, 1975; Albuquerque et al, 1976b;

Engel *et al*, 1977a). Restoration of function in the remaining receptors by reducing antibody levels is apparently sufficient to restore neuromuscular transmission to a level at which the safety factor for transmission can counterbalance the receptor loss and the patient is relatively asymptomatic. This is an important point as it indicates the degree of receptor loss that can occur without clinically significant disease; thus even in acute fulminating cases substantial endplate destruction and receptor loss could occur before the development of subjective weakness.

Reduction in Acetylcholine Receptor Number

The cause of the receptor loss in myasthenia is not clear. Two major possibilities exist: that they are destroyed by an early cellular or complement-mediated attack on the endplate during the phase of the disease before the patient presents with weakness; alternatively that the receptor loss results from the type of damage that may be seen following long-term administration of anticholinesterase drugs (Engel *et al*, 1973; Chang *et al*, 1973; Schwartz *et al*, 1977). The first suggestion is supported to some degree by studies on the experimental model of myasthenia (experimental autoimmune myasthenia gravis – EAMG) that has been developed in parallel with the recent studies on the human disease. Indeed it was the observation of the experimental syndrome (Patrick and Lindstrom, 1973) that was largely responsible for the resurgence of interest in human myasthenia as being an autoimmune disorder due to acetylcholine receptor antibody.

Experimental Autoimmune Myasthenia Gravis

EAMG is produced by immunising animals with acetylcholine receptor preparations, often from electric eel or torpedo electric organs, but also using syngeneic acetylcholine receptor (Patrick and Lindstrom, 1973; Green *et al*, 1975; Tarrab-Hazdai *et al*, 1975b; Lennon *et al*, 1975; Lindstrom *et al*, 1976a, b). In parallel with a rise in antibody levels against receptor the animals develop weakness that resembles myasthenia in many of its clinical and electrophysiological features. It has two phases, an acute phase in which cellular infiltration and anti-body-dependent damage of the endplate occur, and a chronic phase which more closely resembles the disease seen in man. An acute disease may be effected by passive transfer of IgG from immunised animals (Lindstrom *et al*, 1976b). Ultrastructural and biochemical studies of animals with EAMG show striking similarities to the changes seen in human disease in terms of the reduction in postsynaptic surface and the number of acetylcholine receptors; in both the human disease and the animal model the severity of weakness correlates closely with the amount of antibody bound to the remaining receptors (Engel *et al*, 1977; Lindstrom and Lambert, 1978). Thus while the initiation of autoimmune injury in EAMG by immunisation is clearly artificial, the extensive similarities suggest that human myasthenia gravis may be largely, if not all, explicable in terms of autoimmune attack or blockade of acetylcholine receptors mediated by specific antibody.

Autoantibody, the Thymus and Viruses

An important, and so far largely unanswered, question is what provokes the disease in man. At least a proportion of patients with myasthenia gravis have an autoimmune predisposition, in that a strong association with other autoimmune disorders is seen (e.g. thyroiditis, pernicious anaemia, insulin-dependent diabetes). Myasthenia affecting the younger age group shares an HLA association (B 8) with these diseases. This or closely adjacent genes may confer a susceptibility to the formation of autoantibodies, possibly initiated by virus-induced exposure of self antigens or loss of tolerance. The occurrence of thymic abnormalities in this group and the high frequency of thymomas in the older (non-HLA associated) group of myasthenics suggests that the antigen may be derived from the thymus, with cross-reactivity against acetylcholine receptor. The thymus is known to contain muscle-like or myoid cells and these have been shown to have acetylcholine receptor-like material capable of binding α-bungarotoxin (Aharonov *et al*, 1975b; Lindstrom *et al*, 1976d; Kao and Drachman, 1977a; Engel *et al*, 1977c). The exposure of acetylcholine receptor-like antigenic determinants following viral thymitis or following malignant transformation in the thymus is an attractive possibility as the immunising event in human myasthenia gravis. The recent demonstration of raised CMV titres in patients with myasthenia before the use of steroids or thymectomy suggests that CMV could be the viral agent responsible (Tindall *et al*, 1978).

Cell-Mediated Immunity

Cell-mediated immunity has also been implicated in myasthenia gravis, with evidence from lymphocyte stimulation studies suggesting cell mediated immunity to acetylcholine receptor (Abramsky *et al*, 1975, Richman *et al*, 1976), but the evidence is subject to the limitations implicit in investigation of this aspect of immunity. However, passive transfer of the experimental disease by lymph node cells has been achieved (Tarrab-Hazdai *et al*, 1975a). On current evidence it is likely that, despite the known involvement of the thymus in cell mediated immunity, thymic disorders in myasthenia are related to other aspects of thymic structure and function. Some recent evidence has suggested that the synthesis of acetylcholine receptor antibody may, at least in part, occur in the thymus (Vincent *et al,* 1978).

Congenital Myasthenia

Patients with a rare form of myasthenia gravis, congenital myasthenia, have the disease from birth and may be the children of consanguinous marriages; they do not have circulating acetylcholine receptor antibody (Ito *et al,* 1978a). A single patient with this variety has been treated with plasma-exchange and showed no benefit, confirming that this type of the disease is not mediated by humoral factors (Pinching *et al,* 1976; Newsom Davis *et al,* 1978) and lending further support to the idea that the response of patients with acquired myasthenia gravis is related to the removal of acetylcholine receptor antibody. It is probable that in congenital myasthenia the acetylcholine receptor is structurally abnormal.

The Eaton-Lambert Syndrome

By contrast with myasthenia, although the pathophysiological mechanism in the Eaton-Lambert syndrome seems to have been identified, its precise pathogenesis is far from clear. Ultrastructural studies (Engel and Santa, 1971) show increased infolding of the endplate region. The m.e.p.p.'s are normal in amplitude and their frequency is normal or slightly increased. However the release of acetylcholine in response to nerve stimulation is much reduced (10% of normal), resulting in a reduced endplate potential (Lambert and Elmqvist, 1971); this may be inadequate to reach the threshold for depolarisation. The acetylcholine receptor content of the neuromuscular junction is normal and there are no acetylcholine receptor antibodies, either in serum or at the endplate (Lindstrom and Lambert, 1978). So the defect in the Eaton-Lambert syndrome is one of acetylcholine release following nerve stimulation.

Although the cause of this is not clear, recent work (Ishikawa *et al,* 1977) has shown that an extract of the tumour from a patient with the syndrome was able to decrease the endplate potential of a frog neuromuscular preparation under suitable conditions. The endplate potential could be restored by washing. A similar extract from a tumour not associated with the syndrome had no effect. It is notable however that prolonged washing of the affected human neuromuscular junctions in Lambert and Elmqvist's study (1971) did not restore the endplate potentials.

Nevertheless this work does suggest that there is a circulating tumour-associated substance which has the capacity to interfere with acetylcholine release. Its identification would be of considerable interest, not least because it might provide a means for studying the mechanism of co-ordinated acetylcholine release from the nerve terminal after stimulation; the fact that spontaneous release of quanta is unaffected in the Eaton-Lambert syndrome suggests that the mechanisms of spontaneous and co-ordinated release are different.

Conclusions

Recent investigations into the electrophysiology and immuno-pathogenesis of myasthenia gravis and to a lesser extent of the Eaton-Lambert syndrome have provided a number of concepts which may be applied more widely to neurosecretory and indeed hormonal systems. It is possible that comparable blocking factors or specific defects either of transmitter release or of receptor function may underlie several central nervous system diseases. The specific disorders of the neuromuscular junction may serve as a model for a variety of cell to cell interactions involving the mediation of chemical agents and membrane receptor molecules.

Acknowledgements

I would like to thank Dr. J. Newsom Davis, Dr. A. Vincent and Professor D.K. Peters for many valuable discussions concerning these disorders.

References

Abramsky, O., Aharonov, A., Webb, C. and Fuchs, S. (1975). *Clin. exp. Immunol.* **19**, 11–16.

Aharonov, A., Abramsky, O., Tarrab-Hazdai, R. and Fuchs, S. (1975a). *Lancet* **ii**, 340–342.

Aharonov, A., Tarrab-Hazdai, R., Abramsky, O. and Fuchs, S. (1975b). *Proc. Nat. Acad. Sci. (USA)* **72**, 1456–1459.

Albuquerque, E.X., Rash, J.E., Mayer, R.F. and Satterfield, J.R. (1976a). *Exp. Neurol* **51**, 536–563.

Albuquerque, E.X., Lebeda, F.J., Appel, S.H., Almon, R., Kaufmann, F.C., Mayer, R.F., Narahashi, T. and Yeh, J.Z. (1976b). *Am. N.Y. Acad. Sci.* **274**, 475–492.

Almon, R.R., Andrew, C.G. and Appel, S.H. (1974). *Science* **186**, 55–57.

Almon, R.R. and Appel, S.H. (1975). *Biochemica et Biophysica Acta* **393**, 66–77.

Anwyl, R., Appel, S.H. and Narahashi, T. (1977). *Nature* **267**, 262–263.

Appel, S.H., Almon, R.R. and Levy, N. (1975). *N.E.J.M.* **293**, 760–761.

Appel, S.H., Anwyl, R., McAdams, M.W. and Elias, S. (1977). *Proc. Nat. Acad. Sci. (USA)* **74**, 2130–2134.

Bender, A.N., Ringel, S.P., Engel, W.K., Daniels, M.P. and Vogel, Z. (1975). *Lancet* **i**, 607–608.

Bergstrom, K., Franksson, C., Matell, G. and von Reis, G. (1973). *Europ. Neurol.* **9**, 157–167.

Bergstrom, K., Franksson, C., Matell, G., Nilsson, B.Y., Persson, A., von Reis, G. and Stensman, R. (1975). *Europ. Neurol.* **13**, 19–30.

Bevan, S., Kullberg, R.W. and Heinemann, S.F. (1977). *Nature* **267**, 263–265.

Chang, C.C., Chen, T.F. and Chuang, S-T. (1973). *J. Physiol.* **230**, 613–618.

Dau, P.C., Lindstrom, J.M., Cassell, C.K., Denys, E.H., Shev, E.E. and Spitler, L.E. (1977). *N.E.J.M.* **297** 1134–1140.

Drachman, D.B. (1978). *N.E.J.M.* **298**, 136–142, 186–193.

Elmqvist, D., Hofmann, W.W., Kugelberg, J. and Quastel, D.M.J. (1964). *J. Physiol.* **174**, 417–434.

Engel, A.G. and Santa, T. (1971). *Ann. NY Acad. Sci.* **183**, 46–63.

Engel, A.G., Lambert, E.H. and Santa, T. (1973). *Neurology (Minn).* **23**, 1273–1281.

Engel, A.G., Lindstrom, J.M., Lambert, E.H. and Lennon, V.A. (1977a). *Neurology (Minn).* **27**, 307–315.

Engel, A.G., Lambert, E.H. and Howard, F.M. (1977b). *Mayo Clin. Proc.* **52**, 267–280.

Engel, W.K., Trotter, J.L., McFarlin, D.E. and McIntosh, C.L. (1977c). *Lancet* **i**, 1310–1311.

Fambrough, D.M., Drachman, D.B. and Satyamurti. (1973). *Science* **182**, 293–295.

Green, D.P.L., Miledi, R. and Vincent, A. (1975). *Proc. Roy. Soc. (Lond) B.* **189**, 57–68.

Ishikawa, K., Engelhardt, J.K., Fujisawa, T., Okamoto, T. and Katsuki, H. (1977). *Neurol. (Minn).* **27**, 140–143.

Ito, Y., Miledi, R., Molenaar, P.C., Vincent, A., Polak, R.L., van Gelder, M. and Newsom Davis, J. (1976). *Proc. R. Soc. (Lond.) B.* **192**, 475–480.

Ito, Y., Miledi, R., Molenaar, P.C., Newsom Davis, J., Polak, R.L. and Vincent, A. (1978a). *In* "Biochemistry of Myasthenia Gravis and Muscular Dystrophy" (eds. Marchbanks, R. and Lunt, G.), pp. 89–109. Academic Press, London & New York.

Ito, Y., Miledi, R., Vincent, A. and Newsom Davis, J. (1978b). *Brain* **101** (In Press).

Kao, I. and Drachman, D.B. (1977a). *Science* **195**, 74–75.

Kao, I. and Drachman, D.B. (1977b). *Science* **196**, 527–529.

Keesey, J., Lindstrom, J., Cokely, H. and Hermann, C. (1977). *N.E.J.M.* **296**, 55.

Lambert, E.H. and Elmqvist, D. (1971). *Ann. N.Y. Acad. Sci.* **183**, 183–199.

Lefvert, A.K. and Bergstrom, K. (1976). *Europ. J. Clin. Invest.* **7**, 115–119.

Lennon, V.A., Lindstrom, J.M. and Seybold, M.E. (1975). *J. Exp. Med.* **141**, 1365–1375.

Lindstrom, J.M. and Lambert, E.H. (1978). *Neurology (Minn).* **28**, 130–138.

Lindstrom, J.M., Einarson, B.L., Lennon, V.A. and Seybold, M.E. (1976a). *J. Exp. Med.* **144**, 726–738.

Lindstrom, J.M., Engel, A.G., Seybold, M.E., Lennon, V.A. and Lambert, E.H. (1976b). *J. Exp. Med.* **144**, 739–753.

Lindstrom, J.M., Seybold, M.E., Lennon, V.A., Whittingham, S. and Duane, D.D. (1976c). *Neurology (Minn).* **26**, 1054–1059.

Lindstrom, J.M., Lennon, V.A., Seybold, M.E., and Whittingham, S. (1976d). *Ann. N.Y. Acad. Sci.* **274**, 254–274.

Mittag, T., Kornfeld, P., Tormay, A. and Woo, C. (1976). *N.E.J.M.* **294**, 691–694.

Namba, T., Brown, S.B. and Grob, D. (1970). *Pediatrics* **45**, 488–504.

Newsom Davis, J., Pinching, A.J., Vincent, A. and Wilson, S.G. (1978). *Neurology (Minn).* **28**, 266–277.

Patrick, J. and Lindstrom, J.M. (1973). *Science* **180**, 871–872.

Pinching, A.J. (1978). *Br. J. Hosp. Med.* (In Press).

Pinching, A.J., Peters, D.K. and Newsom Davis, J. (1976). *Lancet* ii, 1373–1376.

Rash, J.E., Albuquerque, E.X., Hudson, C.S., Mayer, R.F. and Satterfield, J.R. (1976). *Proc. Nat. Acad. Sci. USA* **73**, 4584–4588.

Richman, D.P., Patrick, J. and Arnason, B.G.W. (1976). *N.E.J.M.* **294**, 694–698.

Schwartz, M.S., Sargeant, M.K. and Swash, M. (1977). *Neurology* **27**, 289–293.

Simpson, J.A. (1960). *Scot. Med. J.* **5**, 419–436.

Tarrab-Hazdai, R., Aharonov, A., Abramsky, O., Yaar, I. and Fuchs, S. (1975a). *J. Exp. Med.* **142**, 785–789.

Tarrab-Hazdai, R., Aharonov, A., Silman, I., Fuchs, S. and Abramsky, O. (1975b). *Nature* **256**, 128–130.

Tindall, R.S.A., Cloud, R., Luby, J. and Rosenberg, R.N. (1978). *Neurology (Minn).* **28**, 273–277.

Toyka, K.V., Drachman, D.B., Griffin, D.E., Pestronk, A., Winkelstein, J.A., Fischbeck, K.H. and Kao, I. (1977). *N.E.J.M.* **296**, 125–131.

Vincent, A., Scadding, G.K., Thomas, H.C. and Newsom Davis, J. (1978). *Lancet* i, 305–307.

MODERN CONCEPTS OF ADRENERGIC TRANSMISSION

S.Z. LANGER

*Laboratoires d'Etudes et de Recherches Scientifiques,
Department of Biology, Paris, France.*

Introduction

A great deal of progress has taken place since the discovery by von
Euler (1946) that the neurotransmitter released from adrenergic nerve
terminals was noradrenaline. The terminal varicosities of the
noradrenergic neurons are involved in the synthesis, storage, release
and inactivation of the neurotransmitter. The transmitter released
upon the arrival of nerve impulses activates specific adrenoceptors
located in the membrane of the postsynaptic effector cell (in the
periphery) or of the postsynaptic neuron (in the central nervous
system).

   Until a few years ago the noradrenergic nerve terminals were
thought to be devoid of receptor sites, and it was generally accepted
that adrenoceptors were located solely on the membrane of postsynaptic structures. It is only during the past few years that evidence
has become available for the presence of adrenoceptors and other
receptor systems on the outer surface of the membrane of the
noradrenergic varicosities (Langer, 1974a, 1977; Starke, 1977). These
receptors, which are called presynaptic because of their location, are
involved in the modulation of stimulation-evoked transmitter
release.

   There are several target mechanisms for the action of drugs on
noradrenergic neurotransmission: a) synthesis; b) storage; c) release
— either through nerve impulses or by displacing agents; d) presynaptic
receptors that modulate transmitter release; e) postsynaptic
adrenoceptors which mediate the responses; and f) inactivating
mechanisms — neuronal uptake, extraneuronal uptake and the
catabolising enzymes monoamine oxidase and catechol-O-
methyltransferase. Only the most recent developments in certain
areas will be discussed in some detail in the present chapter.

Synthesis

The precursor of noradrenaline, tyrosine, is present in the circulation and in all tissues. Tyrosine enters the neurone through a mechanism which differs from the active transport site present for noradrenaline in the nerve terminals. Tyrosine is acted upon by tyrosine hydroxylase, a soluble enzyme, to form dihydroxyphenylalanine (DOPA). This reaction is the rate-limiting step in the synthesis of the neurotransmitter, and the enzyme, tyrosine hydroxylase, can be inhibited by the end product, noradrenaline, and by other catechols as well, like dopamine and DOPEG (3, 4-dihydroxyphenylglycol) the deaminated catabolite of noradrenaline which is formed preferentially in the nerve ending (Langer, 1974b). Decarboxylation of DOPA to form dopamine also occurs in the cytoplasm, and this enzymatic step is not rate-limiting. Dopamine is the immediate precursor of noradrenaline, and it enters the vesicular storage site to be beta-hydroxylated by dopamine-beta-hydroxylase, an intravesicular enzyme. It is of interest to note that the endogenous levels of dopamine in noradrenergically innervated tissues of the peripheral nervous system do not exceed 5% of the total level of catecholamines (Bell *et al*, 1978).

These results indicate that under physiological conditions dopamine does not accumulate in noradrenergic nerve endings. Consequently the hypothetical release of small amounts of dopamine together with noradrenaline during nerve activity does not seem to be important under physiological conditions.

In the adrenal medulla and in certain neurons of the central nervous system there is an additional synthetic step, catalysed by the enzyme phenylethanolamine-N-methyl transferase (PNMT). Noradrenaline is thus N-methylated to adrenaline in the adrenal medulla, which is the main source of circulating adrenaline.

In the central nervous system the areas rich in PNMT have been found to contain adrenaline, which appears to subserve the function of a neurotransmitter (Saavedra *et al*, 1974; Hokfelt *et al*, 1974). Some of these regions, like the hypothalamus and the nucleus tractus solitari, are known to be involved in the regulation of blood pressure.

Storage

Noradrenaline is stored in nerve endings, probably complexed with ATP in the vesicular storage sites from which it can be released upon the arrival of nerve impulses.

Reserpine is the classical drug that interferes with the storage of the neurotransmitter and leads to a depletion of the endogenous stores of noradrenaline in the peripheral and the central nervous system. The administration of reserpine also depletes the endogenous levels of dopamine and 5-hydroxytryptamine in the central nervous system. During reserpine-induced depletion, noradrenaline is metabolized in the cytoplasm of the nerve endings by monoamine oxidase, and it is therefore lost in the form of deaminated metabolites. It is of interest to note that administration of reserpine induces a total depletion of the endogenous levels of noradrenaline without affecting the vesicular content of dopamine-beta-hydroxylase (Cubeddu and Weiner, 1975).

Although reserpine has been mainly used clinically in the treatment of hypertension it is of interest to note that sedation and occasionally a depressive syndrome are among its side effects.

Release of Noradrenaline: Role of Presynaptic Receptors

The physiological mechanism for the release of noradrenaline upon the arrival of nerve impulses involves a calcium-dependent exocytotic process. The transmitter released by nerve impulses activates post-synaptic adrenoceptors to elicit the classical response of the effector organ (in the periphery) or of the postsynaptic neuron (in the central nervous system).

Noradrenaline can also be released by indirectly acting sympathomimetic amines like tyramine. This release is due to the displacement of noradrenaline from vesicular storage sites and does not involve exocytosis. In contrast with the release induced by nerve impulses, noradrenaline release induced by tyramine is not calcium-dependent and is not accompanied by the release of dopamine-beta-hydroxylase.

As mentioned in the Introduction, it is now well accepted that there are receptor sites in the outer surface of noradrenergic nerve endings. These presynaptic receptors are involved in the modulation of the release of noradrenaline during nerve stimulation.

The presence of presynaptic inhibitory alpha-adrenoceptors in the noradrenergic nerve endings of the peripheral nervous system was first reported by Langer *et al* (1971).

These presynaptic alpha-adrenoceptors are involved in the regulation of noradrenaline release through a negative feed-back mechanism mediated by the neurotransmitter itself. Consequently, noradrenaline released by nerve stimulation, once it reaches a threshold concentration in the synaptic cleft, activates presynaptic alpha-adrenoceptors, triggering a negative feed-back mechanism which inhibits further release of the neurotransmitter (for reviews see Langer 1974a, 1977; Starke, 1977).

In support of this hypothesis it has been demonstrated that alpha-adrenoceptor agonists reduce noradrenaline release during nerve stimulation, independently of the alpha or beta nature of the post-synaptic adrenoceptor that mediates the response of the effector organ.

The magnitude of the reduction in stimulation-evoked release of noradrenaline induced by alpha-adrenoceptor agonists is more pronounced the lower the frequency of nerve stimulation (Dubocovich and Langer 1974, 1976; Langer *et al*, 1975a). In fact, alpha-adrenoceptor agonists fail to reduce the output of noradrenaline induced by high frequency nerve stimulation. The latter is in contrast with the effects of neuron blocking agents like guanethidine or bretylium, which decrease the release of the neurotransmitter at both low and high frequencies of nerve stimulation (Armstrong and Boura, 1973).

Activation of presynaptic alpha-adrenoceptors decreases noradrenaline release by reducing the availability of calcium for the excitation-secretion coupling (Langer *et al*, 1975a). It is of interest to note that the mechanism that regulates noradrenaline release is operative for the release induced by nerve stimulation or by potassium, but not the release elicited by tyramine (Pelayo *et al*, 1978). It is well known that in contrast with the release of noradrenaline elicited by nerve stimulation or by potassium, tyramine releases the neurotransmitter by displacement from vesicular storage sites through a calcium-independent mechanism.

Additional support for the view that noradrenaline release is regulated through a negative feed-back mechanism mediated by presynaptic alpha-adrenoceptors was obtained in experiments in which a marked increase in stimulation-evoked transmitter release was obtained by alpha-adrenoceptor blocking agents (for reviews, see Langer, 1974a, 1977; Starke, 1977).

In support of the view that the alpha-adrenoceptor involved in the negative feed-back mechanism for noradrenaline release is located on the noradrenergic nerve terminals it has been shown that in recently formed nerve endings from cultured rat superior cervical ganglia the alpha-blocking agent phenoxybenzamine enhances the release of ^{3}H-noradrenaline evoked by potassium depolarization (Vogel *et al*, 1972).

More recently it has been demonstrated that the presynaptic regulation of noradrenaline release mediated by alpha-adrenoceptors is not affected after atrophy of the postsynaptic effector cell in the rat salivary gland (Filinger *et al*, 1978). In these salivary glands, which are atrophied as a result of duct ligation, the innervation remains intact but the secretory responses to adrenoceptor or cholinoceptor agonists are abolished. Under these experimental conditions, exposure to phentolamine produced a 3-fold increase in the K^+-evoked release of ^{3}H-noradrenaline from slices of both normal and atrophied salivary gland (Filinger *et al*, 1978). These results indicate that the effects of phentolamine were mediated at a presynaptic site, because they could be demonstrated independently of the presence of a postsynaptic effector cell.

Finally, a significant decrease in the number of binding sites for ^{3}H-dihydroergocryptine was found in the rat heart after degeneration of the noradrenergic nerve endings following the administration of 6-hydroxydopamine (Story *et al*, 1978).

Although both the pre- and the postsynaptic alpha-adrenoceptors are stimulated by alpha-receptor agonists and blocked by alpha-adrenoceptor antagonists, evidence has accumulated during the last few years to support the view put forward by Langer (1973), that the postsynaptic alpha-adrenoceptors which mediate the response of the effector organ are not identical with the presynaptic alpha-adrenoceptors which regulate the release of noradrenaline during nerve stimulation.

Differences in the affinity for agonists and antagonists between the presynaptic and postsynaptic alpha-adrenoceptors are shown in Table 1. These differences have led to the concept (Langer, 1974a) that there are indeed two types of alpha-adrenoceptors, the postsynaptic receptors which mediate on the whole excitatory responses (like vasoconstriction) and which are referred to as α_1, and the presynaptic alpha-receptors which mediate inhibitory effects (reduction of noradrenaline release during nerve stimulation) and which are referred to as α_2.

The affinity for α_1 and α_2 adrenoceptors is similar when agonists like naphazoline, adrenaline and noradrenaline are considered. Tramazoline, alpha-methylnoradrenaline, clonidine and oxymetazoline have a higher affinity for the α_2 adrenoceptor. At the other extreme of the spectrum phenylephrine and methoxamine have a high affinity for the α_1 adrenoceptor and little or no affinity for the presynaptic or α_2 adrenoceptor.

Differences are also present when relative affinities of alpha-blocking agents are considered. Phentolamine has approximately the same potency in blocking the α_1 and the α_2 adrenoceptor. Yohimbine, piperoxan, tolazoline and mianserin are more potent in blockade of the α_2 adrenoceptors than of the α_1 adrenoceptors. Phenoxybenza-

TABLE I

Subclassification of Alpha-Adrenoceptors: Relative Orders of Potency of Agonists and Antagonists

α_1 and α_2 Adrenoceptors	
Relative order of Potency of Agonists	
Tramazoline $>$ αCH_3-Noradrenaline $>$ Clonidine $>$ Oxymetazoline $>$	α_2
$>$ Naphazoline $>$ Adrenaline $>$ Noradrenaline $>$	$\alpha_2 = \alpha_1$
$>$ Phenylephrine $>$ Methoxamine	α_1
Relative Order of Potency of Antagonists	
Yohimbine $\gg$ Piperoxan $>$ Tolazoline $>$ Mianserin	α_2
$>$ Phentolamine $>$	$\alpha_2 = \alpha_1$
$>$ Phenoxybenzamine $\gg$ WB 4101 $\ggg$ Labetolol = Prazosin	α_1

This shows the relative order of potencies of different alpha agonists and antagonists on the postsynaptic (α_1) receptors which mediate the responses of the effector organ and on the presynaptic (α_2) receptors which mediate the inhibition of noradrenaline release during nerve stimulation.
The results shown were obtained in noradrenergically innervated tissues of the peripheral nervous systems of several species.

mine has a higher affinity for the blockade of α_1 adrenoceptors, but as its concentration is increased it will also block the presynaptic or α_2 adrenoceptors (Dubocovich and Langer, 1974). The other three blocking agents, WB 4101, prazosin and labetolol are rather selective for the α_1 adrenoceptor and have a very low or no affinity for the α_2 or presynaptic adrenoceptor. It should be noted that labetolol is also a postsynaptic beta-adrenoceptor blocking agent.

The subclassification shown in Table 1 is valid only for the peripheral nervous system, and as a general guideline. There is increasing evidence of some tissue and species differences between the α_2 adrenoceptors (Roach *et al*, 1978; Arbilla and Langer, 1978a). In the central nervous system it is more difficult to identify α_1 with postsynaptic and α_2 with presynaptic receptors. Yet both types of adrenoceptors are also found in the central nervous system, and their subclassification into α_1 and α_2 follows the relative affinities for agonists and antagonists shown in Table 1.

Changes in sensitivity of the presynaptic alpha-adrenoceptors involved in the regulation of noradrenaline release have been reported recently. Subsensitivity of the presynaptic alpha-adrenoceptors was found in the cat nictitating membrane 18 hours after surgical denervation (Langer and Luchelli-Fortis, 1977). Under these experimental conditions the postsynaptic changes in sensitivity are not yet developed (Langer, 1975) and the subsensitivity of presynaptic alpha-adrenoceptors appears to be due to exposure of the nerve ending to the transmitter leaking from degenerating nerve endings at the onset of the 'degeneration contraction' previously described in this preparation (Langer 1966; Langer and Trendelenburg, 1966).

It is of interest to note that the sensitivity of the presynaptic alpha-adrenoceptor can be reduced after exposure to an effective concentration of an alpha-receptor agonist. In the cat spleen, perfused with cocaine to inhibit neuronal uptake, a short-lasting subsensitivity of both the presynaptic and the postsynaptic alpha-adrenoceptors can be demonstrated after a 60-minute exposure to 0.59 uM of noradrenaline (Langer and Dubocovich, 1977). The short-lasting subsensitivity of the presynaptic alpha-adrenoceptor was reflected by a 2-fold increase in the stimulation-evoked release of ^{3}H-noradrenaline after such a 60-minute exposure.(Langer and Dubocovich, 1977).

These results are compatible with the view that, as already demonstrated for the postsynaptic receptor, chronic stimulation or blockade of the presynaptic receptors may lead to changes in their sensitivity to the neurotransmitter. This phenomenon may contribute to the regulation of neurotransmission, and could be of clinical significance both in the rebound hypertension which occurs after the sudden withdrawal of clonidine and in the mechanism of action of the tricyclic antidepressants, which inhibit neuronal uptake of noradrenaline. Both examples will be discussed in more detail at the end of this chapter.

The physiological significance of the role played by the negative feed-back mechanism mediated by presynaptic alpha-adrenoceptors has been demonstrated *in vitro* through the potentiation of the positive chronotropic responses to cardioaccelerator nerve stimulation, under conditions in which transmitter release was enhanced during exposure to the alpha-receptor blocking agent phentolamine (Langer *et al*, 1977).

Similar results were obtained under *in vivo* experimental conditions: activation of presynaptic alpha-adrenoceptors by clonidine reduces the positive chronotropic responses to cardioaccelerator nerve stimulation (Armstrong and Boura, 1973; Yamaguchi *et al*, 1977) while alpha-receptor antagonists potentiate these positive chronotropic effects (Lokhandwala and Buckley, 1976; Yamaguchi *et al*, 1977).

Evidence is now available for the presence of presynaptic alpha-adrenoceptors in human vasoconstrictor nerves innervating peripheral arteries and veins (Stjarne and Gripe, 1973; Stjarne and Brundin 1975, 1977).

The importance of the presynaptic as well as the postsynaptic effects of phentolamine for the circulatory effects of this alpha-receptor blocking agent in man has recently been demonstrated by Richards et al (1978).

As will be discussed later, neuronal uptake of noradrenaline is the main mechanism for the inactivation of the transmitter released by nerve stimulation. This active transport mechanism for nora-drenaline across the membrane of the nerve ending effectively decreases the concentration of the transmitter in the synaptic gap. It follows, then, that inhibition of neuronal uptake of noradrenaline by cocaine or desipramine will increase that fraction of the nora-drenaline released by nerve stimulation which becomes available for activation of the presynaptic inhibitory alpha-adrenoceptors (Langer, 1974a; Langer and Enero, 1974; Dubocovich and Langer, 1976).

Presynaptic inhibitory alpha-adrenoceptors are also present in the noradrenergic nerve endings of the central nervous system (for reviews see Langer, 1977; Starke, 1977). It appears that the central presynaptic alpha-adrenoceptor resembles the peripheral presynaptic alpha-adrenoceptor in its affinity for agonists as well as for antagonists. Clonidine but not methoxamine reduces the potassium-evoked release of ^{3}H-noradrenaline from slices of the rat occipital cortex (Dubocovich, 1978). On the other hand yohimbine and phentolamine but not prazosin increase the stimulation-evoked release of ^{3}H-noradrenaline from the rat occipital cortex (Dubocovich, 1978).

In addition to the negative feed-back mechanism for noradrenaline released by nerve stimulation, which is mediated by presynaptic alpha-adrenoceptors, there is a positive feed-back mechanism in noradrenergic nerve endings which is triggered through the activation of presynaptic beta-adrenoceptors (Langer et al, 1974; Adler-Graschinsky and Langer, 1975; Langer et al, 1975b; Langer, 1976; Yamaguchi et al, 1977; Celuch et al, 1978; Pelayo et al, 1978; Dahlof et al, 1978). In support of this view it has been reported that exposure to low concentrations of isoprenaline enhances the release of noradrenaline during low-frequency nerve stimulation both in vitro and in vivo. The increase in transmitter release obtained in the presence of beta-adrenoceptor agonists is blocked by low concen-trations of the beta-receptor antagonist, propranolol.

The presence of presynaptic facilitatory beta-adrenoceptors has also been reported in the human oviduct (Hedqvist and Moawad,

1975) and in human vasoconstrictor nerves (Stjarne and Brundin, 1967a, b).

Presynaptic beta-adrenoceptors are present in recently formed noradrenergic nerve endings when rat superior cervical ganglia are cultured (Weinstock *et al*, 1978). Under these experimental conditions the authors are dealing only with nerve endings, since their preparation is devoid of postsynaptic structures.

The facilitatory presynaptic beta-adrenoceptors appear to be of the β_2 rather than β_1 type (Stjarne and Brundin, 1976a; Dubocovich *et al*, 1978). It is of interest to note that very low concentrations of adrenaline can activate presynaptic beta-adrenoceptors, leading to an increase in the stimulation-evoked release of noradrenaline (Stjarne and Brundin, 1975). It is therefore possible that circulating adrenaline is the main stimulus under physiological conditions to trigger the facilitatory mechanism mediated by presynaptic beta-adrenoceptors. The latter may be of particular importance under conditions of stress or when the level of circulating adrenaline is acutely or chronically elevated.

The facilitation of transmitter release triggered by the activation of presynaptic beta-adrenoceptors appears to be mediated through an increase in the levels of cyclic AMP in noradrenergic nerve endings. In support of this view it has been reported that papaverine and other phosphodiesterase inhibitors enhance noradrenaline release during nerve stimulation (Langer, 1973; Cubeddu *et al*, 1975; Celuch *et al*, 1978).

In the perfused cat spleen several cyclic nucleotide analogues enhance both noradrenaline and dopamine-beta-hydroxylase release during nerve stimulation (Cubeddu *et al*, 1975). Similar results are obtained in the guinea-pig heart with low concentrations of dibutyryl c-AMP when the presynaptic alpha-adrenoceptors are blocked with phenoxybenzamine (Langley and Weiner, 1978). More recently it was reported that dibutyryl c-AMP enhances the K^+-evoked release of ^{3}H-noradrenaline from the rat pineal gland (Pelayo *et al*, 1978). These authors also reported that, in contrast to the K^+-evoked release, dibutyryl c-AMP failed to increase the release of ^{3}H-noradrenaline induced by tyramine.

Papaverine induces a shift to the left in the concentration/effect curve for isoprenaline on transmitter release in the perfused cat spleen (Celuch *et al*, 1978). In the rat pineal, when the presynaptic alpha-adrenoceptors are blocked by yohimbine, exposure to a selective cyclic AMP-phosphodiesterase inhibitor produces a 2-fold increase in the K^+-evoked release of ^{3}H-noradrenaline (Pelayo *et al*,

1978).

The possibility that adenylate cyclase present in noradrenergic nerve endings was involved in the activation of tyrosine hydroxylase activity following nerve stimulation was proposed by Roth *et al* (1975). It is not yet clear whether a common presynaptic mechanism is involved both in the facilitation of noradrenaline release mediated by presynaptic beta-adrenoceptors and in the increase in tyrosine hydroxylase activity that occurs as a result of sympathetic nerve stimulation. Both phenomena appear to be mediated through an increase in c-AMP levels in nerve endings.

Although there is good evidence for the presence of presynaptic alpha-adrenoceptors in the central nervous system, the presence there of presynaptic beta-adrenoceptors which facilitate the stimulation-evoked release of noradrenaline is still an open question. Recently, Taube *et al* (1977) reported that isoprenaline failed to increase while propranolol did not decrease the stimulation-evoked release of ^{3}H-noradrenaline from slices of the rat occipital cortex. Since the presynaptic beta-adrenoceptors appear to be triggered when the degree of depolarization of the nerve endings is small or moderate, a more detailed and extensive study would be required to exclude the presence of presynaptic facilitatory beta-adrenoceptors in the central nervous system. Relevant to this point is the recent report by Adler-Graschinsky and Martinez (1978), showing that when ^{3}H-noradrenaline release from the cerebral cortex is elicited by exposure to only 10 mM of potassium there is a significant decrease in the release of the labelled transmitter in the presence of propranolol.

In addition to the presynaptic alpha and beta-adrenoceptors, a dopamine-sensitive inhibitory presynaptic receptor has been described in the noradrenergic nerve endings of several tissues (Langer, 1973; Enero and Langer, 1975; Long *et al*, 1975).

Pharmacological studies using phentolamine to block presynaptic alpha-adrenoceptors have demonstrated that the inhibition of noradrenergic neurotransmission induced by dopamine or apomorphine is not mediated by presynaptic alpha-receptors but rather by pre-synaptic dopamine receptors (Enero and Langer, 1975; Long *et al*, 1975; Langer and Dubocovich, 1978). The reduction in the stimu-lation-evoked release of noradrenaline obtained by dopamine agonists is selectively antagonized by neuroleptics like chlorpromazine, pimozide or sulpiride.

In contrast with the presynaptic alpha-adrenoceptors, which have been found to be present in every tissue in which they have been studied (Langer 1974a, 1977; Starke, 1977) the presynaptic inhibitory dopamine receptors are less widely distributed.

It appears that the presynaptic dopamine receptors which inhibit noradrenaline release during nerve stimulation differ from the post-synaptic vascular dopamine receptor (Goldberg *et al*, 1978). This conclusion is based on differences in affinities for agonists and antagonists between the presynaptic dopamine receptor that inhibits noradrenergic neurotransmission and the postsynaptic or vascular dopamine receptor that causes vasodilation.

The presynaptic dopamine receptors are of pharmacological rather than physiological importance in peripheral noradrenergic neuro-transmission, because selective presynaptic dopamine blocking agents like sulpiride do not increase the stimulation-evoked release of noradrenaline in the way that alpha-adrenoceptor blocking agents do.

In the central nervous system there is as yet no evidence for the presence of inhibitory presynaptic dopamine receptors in nora-drenergic nerve endings. There is however evidence for the presence of presynaptic inhibitory dopamine receptors on the nerve terminals of central dopaminergic neurons (Farnebo and Hamberger, 1971; Plotsky *et al*, 1977; Hertting *et al*, 1978; Nagy *et al*, 1978).

In addition to the presynaptic receptors through which the trans-mitter can regulate its own release, a real mosaic of presynaptic receptors located on noradrenergic nerve endings has been described in recent years. Presynaptic receptors which are involved in the inhibition of noradrenaline release during nerve stimulation include the muscarinic cholinergic receptor, the prostaglandin E series receptor, and in some neurons the opiate and adenosine presynaptic inhibitory receptors (Langer, 1977; Starke, 1977). Another presynaptic receptor, involved in facilitation of the stimulation-evoked release of noradrenaline, is the angiotensin II receptor. This mosaic of presynaptic receptors in peripheral noradrenergic nerve terminals can thus be activated by other neurotransmitters (acetylcholine), circulating endogenous compounds (angiotensin II) or locally produced substances (prostaglandin or adenosine) to modulate transmitter release.

In the central nervous system, too, there is increasing evidence for modulation of the release of neurotransmitters through presynaptic receptor mechanisms. In the rat cerebral cortex morphine and β-endorphin reduce the stimulation-evoked release of ^{3}H-noradrenaline through a naloxone-sensitive mechanism (Arbilla and Langer, 1978b). On the other hand GABA enhances the stimulation-evoked release of ^{3}H-noradrenaline from the rat occipital cortex, but not from the peripheral noradrenergic nerve endings of the cat nictitating membrane (Arbilla and Langer, 1978c). Thus local neurotransmitter interactions may represent a physiological mechanism for the modulation of transmitter release in the central nervous system.

Attention should be drawn to the fact that presynaptic receptors can also be acted upon by exogenously administered agonists or antagonists to modulate the release of neurotransmitters induced by nerve impulses. In view of the differences that have been reported between the presynaptic and the postsynaptic receptors (alpha-adrenergic and dopamine receptors) it is possible that selective agonists or antagonists might elicit pharmacological effects which result from changes in transmitter release. The possible therapeutic potential and some of the clinical implications are discussed below.

Effects on Postsynaptic Adrenoceptors

The physiological effects of noradrenaline released from nerves are mediated through the classical postsynaptic receptors: α_1 and β_1 and β_2 adrenoceptors. While these receptors are well defined in the periphery and the effects mediated by their activation are well known, the situation is quite different in the central nervous system. A few examples will be discussed to illustrate this point.

Clonidine is an alpha-adrenoceptor agonist, and it can produce vasoconstriction when injected intravenously. However, in the central nervous system clonidine stimulates selectively inhibitory alpha-adrenoceptors, resulting in a decrease in heart rate and blood pressure. Clonidine decreases the turnover of noradrenaline and adrenaline in the rat brain (Scatton et al, 1978) and this effect is selectively antagonized by alpha-receptor blocking agents like yohimbine. The central alpha-adrenoceptor involved in the cardio-vascular inhibitory effects of clonidine appears to be postsynaptic, because the intracisternal administration of clonidine reduces the spontaneous electrical activity of splanchnic nerve fibres after depletion of their endogenous stores of neurotransmitter by the simultaneous administration of reserpine and alpha-methyl-p-tyrosine (Kobinger, 1978). In addition, the sedative effects of clonidine appear to be related to the stimulation of central receptors resembling peripheral presynaptic alpha-adrenoceptors (Delbarre and Schmitt, 1971; Drew et al, 1977). However, the central alpha-adrenoceptors responsible for the antihypertensive effects of clonidine as well as those mediating sedation appear to have the α_2 adrenoceptor properties as far as affinities for agonists and antagonists are concerned (Table 1).

Classical alpha-receptor blocking agents like phentolamine are not useful in the treatment of hypertension, and furthermore they produce marked tachycardia. On the other hand, prazosin, which is a selective postsynaptic alpha-receptor blocking agent with no affinity for presynaptic alpha-adrenoceptors (Table 1) is effective in lowering blood pressure in hypertensive patients, and this effect is not followed by tachycardia. It is possible that this difference between the cardiovascular effects of phentolamine and prazosin is at least partly due to the fact that phentolamine is equally effective in blocking both presynaptic and postsynaptic alpha-adrenoceptors (Table 1).

Alpha-adrenoceptor blocking agents have no predictive antidepressant activity. Yet, a new antidepressant, mianserin, blocks presynaptic alpha-receptors in the central nervous system and thus increases the stimulation-evoked release of noradrenaline (Baumann and Maitre, 1977). Yohimbine blocks preferentially presynaptic alpha-adrenoceptors (Table 1) and preliminary reports indicate that this agent has an antidepressant effect on man (Puech *et al*, 1978). Furthermore, clonidine, an alpha-receptor agonist with selectivity for presynaptic receptors (Table 1) can produce depression in man in the course of treatment for hypertension (Puech *et al*, 1978). These results taken together are compatible with the view that presynaptic alpha-adrenoceptors in the central nervous system might be involved in mood regulation through their effects on noradrenaline release.

Recently it was reported that salbutamol, a β_2-adrenoceptor agonist, showed a clear antidepressant effect in most patients with severe depressive states (Widlocher *et al*, 1977). Of special interest is the fact that the onset of this antidepressant effect appeared earlier than that of clomipramine (Puech *et al*, 1978). The possible role of central beta-adrenoceptors in the mechanism of action of antidepressant drugs is still unclear. However, it should be noted that chronic treatment with antidepressant drugs results in subsensitivity of the noradrenergic receptor coupled adenylate cyclase system in the limbic forebrain (Sulser *et al*, 1978). In addition Banerjee *et al* (1977) have reported that chronic treatment with several antidepressants resulted in subsensitivity of beta-adrenoceptors. These authors found a decrease in ^{3}H-dihydroalprenolol binding in the rat brain after chronic desipramine treatment (Banerjee *et al*, 1977).

Inactivating Mechanisms

Neuronal uptake of noradrenaline is the main mechanism for terminating the action of neurotransmitter released during nerve stimulation. This active transport mechanism across the membrane of the nerve terminal is also the main inactivating mechanism for exogenously administered sympathomimetic amines, which are a good substrate for the uptake system (Trendelenburg, 1963).

Inhibition of neuronal uptake of noradrenaline by cocaine or desipramine results in a potentiation of the responses of the effector organ both to exogenous noradrenaline (Langer and Trendelenburg, 1969) and to sympathetic nerve stimulation (Langer and Enero, 1974). The potentiation of the responses to exogenous noradrenaline and to nerve stimulation is causally related to the increase in concentration of noradrenaline in the synaptic cleft as a result of the inhibition of neuronal uptake.

Access of noradrenaline to the intraneuronal metabolizing enzymes is mediated through the neuronal uptake mechanism. The transmitter taken up in the noradrenergic nerve terminals is either stored in the vesicles or deaminated by monoamine oxidase and subsequently reduced by aldehyde reductase (Langer, 1974b). As a result, the main metabolite of noradrenaline formed within the nerve terminal is the deaminated glycol DOPEG (3, 4-dihydroxyphenylglycol). It is of interest that DOPEG formation represents the first step in the catabolism of noradrenaline in both the peripheral and the central nervous systems (Langer 1974b; Langer and Enero, 1974; Luchelli-Fortis and Langer, 1975; Dubocovich and Langer, 1976; Farah *et al*, 1977). In the central nervous system the main final product of the catabolism of noradrenaline is represented by the O-methylation product of DOPEG, 3-methoxy-4-hydroxyphenylglycol (MOPEG), which is present in both the free and the conjugated form. MOPEG sulphate is present in the cerebro-spinal fluid, and its levels appear to be correlated with central noradrenergic activity.

There is a second uptake mechanism at the noradrenergic neuroeffector junction. This is the extraneuronal uptake mechanism, which represents the access of noradrenaline to the postsynaptic catabolising enzymes, particularly catechol-O-methyltransferase (COMT). This extraneuronal uptake mechanism can be blocked by metanephrine, normetanephrine or hydrocortisone. Extraneuronal uptake coupled with O-methylation represents an inactivating pathway for sympathomimetic amines like isoprenaline which are not taken up by the neuronal uptake mechanism. Potentiation of the responses of the effector organ to isoprenaline can be obtained when either COMT or extraneuronal uptake are inhibited.

In the peripheral nervous system inhibition of either monoamine oxidase or COMT or both simultaneously does not potentiate the responses of the effector organ to endogenously released or exogenously administered noradrenaline (Langer and Enero, 1974). In the central nervous system, inhibition of monoamine oxidase is associated with anti-depressant activity.

Possible Pharmacological and Clinical Significance of Presynaptic Receptor Systems

Presynaptic Alpha-adrenoceptors

Clonidine is an alpha-adrenoceptor agonist which produces its anti-hypertensive effects through the activation of central alpha-adrenoceptors probably located postsynaptically (Kobinger, 1978). As pointed out earlier clonidine decreases the turnover of both noradrenaline and adrenaline in the rat brain under *in vivo* conditions, and this effect is antagonized by selective α_2-adrenoceptor blocking agents like yohimbine (Scatton *et al*, 1978). In addition the administration of clonidine results in a decrease in the brain levels of the main catabolite of noradrenaline, MOPEG (Braestrup and Nielsen, 1976). It is possible that the effects of clonidine on the turnover of noradrenaline and adrenaline and on MOPEG levels are causally related to the activation of central presynaptic alpha-adrenoceptors which decrease the release of the neurotransmitter. As already pointed out, the sedative effects of clonidine appear to be related to the stimulation of central receptors which are very similar to the peripheral presynaptic alpha-adrenoceptors (Delbarre and Schmitt, 1971; Drew *et al*, 1977).

Clonidine activates presynaptic receptors in the peripheral nervous system, and its affinity for the presynaptic or α_2 receptor is considerably higher than that for the postsynaptic α_1 adrenoceptor (Starke *et al*, 1974). In fact *in vitro* concentrations as low as 0.01 nM of clonidine result in a significant decrease of the stimulation-evoked release of noradrenaline (Medgett *et al*, 1978). Consequently, it is possible that even low plasma levels of clonidine might activate presynaptic alpha-adrenoceptors in the peripheral nervous system, leading to an additional decrease in sympathetic tone which can contribute to the whole antihypertensive and especially to the bradycardic action of this drug.

As discussed above, subsensitivity of the presynaptic alpha-adrenoceptor has been reported under various experimental conditions (Langer and Dubocovich, 1977; Langer and Luchelli-Fortis, 1977). As reported by these authors, such subsensitivity is reflected by an increase in the stimulation-evoked release of noradrenaline.

It is possible that after chronic treatment with clonidine some degree of subsensitivity may develop in the presynaptic alpha-adrenoceptors, and the hypertensive crisis which results from the sudden interruption of the administration of clonidine may result at least partially from an enhanced release of noradrenaline from sympathetic nerves, due to such subsensitivity.

In support of this view, it has been reported that the alpha-receptor blocking agent phenoxybenzamine becomes less effective in increasing the stimulation-evoked release of noradrenaline after chronic administration of clonidine in the rat (D. Story, personal communication).

Mianserin is an antidepressant agent which blocks presynaptic alpha-adrenoceptors and enhances the stimulation-evoked release of noradrenaline in the central nervous system (Baumann and Maitre, 1977). Since depletion of the noradrenaline stores (by reserpine) or the reduction of stimulation-evoked release of noradrenaline (by clonidine) can be associated with depressive syndromes, it is possible that the increase in noradrenaline release resulting from the blockade of presynaptic alpha-adrenoceptors by mianserin may be related to its clinical antidepressant effects.

The main group of drugs employed in the treatment of depression are the tricyclic derivatives like imipramine, desipramine and its analogues. These drugs are potent inhibitors of the neuronal uptake of noradrenaline. As discussed earlier, inhibition of neuronal uptake of noradrenaline increases the concentration of the neurotransmitter in the synaptic gap. It is therefore possible that chronic inhibition of neuronal uptake or noradrenaline may lead to subsensitivity of the presynaptic alpha-adrenoceptors, as a result of the continuous activation of these receptors resulting from a long-lasting increase in the concentration of noradrenaline in the synaptic cleft (Langer, 1974a; Langer and Dubocovich, 1977; Langer and Luchelli-Fortis, 1977). Consequently it is possible that the chronic administration of the tricyclic antidepressants which inhibit neuronal uptake of noradrenaline may increase the neurally mediated release of the transmitter as a result of the development of subsensitivity of the presynaptic alpha-adrenoceptors. This hypothesis for the mechanism of action of the tricyclic antidepressants is compatible with the latency period which is required for the appearance of the clinical antidepressant effects of these compounds. A certain time is required for the development of subsensitivity of the presynaptic alpha-adrenoceptors which regulate the release of noradrenaline.

Compatible with this hypothesis is the fact that after chronic but not acute administration of desipramine there is an enhancement in the stimulation-evoked release of noradrenaline from the rat heart (Crews and Smith, 1978). This effect is due to the development of subsensitivity in the presynaptic alpha-adrenoceptors of the heart, because phenoxybenzamine produces a 3-fold shift to the left in the frequency response curve to nerve stimulation one day after desipramine treatment but fails to so after three weeks of desipramine treatment (Crews and Smith, 1978).

The development of postsynaptic subsensitivity of the central adrenoceptor after the chronic administration of different types of antidepressants has been reported by Vetulani *et al* (1976). These authors postulated that the reduction in sensitivity of the cyclic AMP generating system to noradrenaline in the limbic forebrain of the rat might be a general pattern for the effects of antidepressants.

As shown by Langer and Dubocovich (1977) subsensitivity of presynaptic alpha-adrenoceptors develops in parallel with subsensitivity of postsynaptic alpha-receptors in the cat spleen. It is therefore tempting to speculate that the subsensitivity of postsynaptic central noradrenergic receptors reported to develop after chronic treatment with antidepressants (Sulser *et al*, 1978) occurs also for the pre-synaptic alpha-adrenoceptor that regulates the release of noradrenaline.

An increase in the neurally mediated release of noradrenaline in the central nervous system resulting either from the blockade or from subsensitivity of presynaptic alpha-adrenoceptors might represent a common mechanism for the antidepressant effects of several drugs. Such a mechanism of action would be compatible with the catecholamine hypothesis of affective disorders as proposed by Schildkraut and Kety (1967).

Presynaptic beta-adrenoceptors

Beta adrenoceptor blocking agents are employed extensively in anti-hypertensive therapy, although the mechanism of their action is not yet clarified. Several authors have suggested that blockade of the presynaptic facilitatory beta-adrenoceptors in the peripheral nervous system may at least contribute to their antihypertensive effects (Adler-Graschinsky and Langer, 1975; Langer *et al*, 1975b; Ljung *et al*, 1975; Langer, 1976; Yamaguchi *et al*, 1977).

Relevant to this proposal is the fact that very low concentrations of adrenaline can increase stimulation-evoked release of noradrenaline through the activation of presynaptic beta-adrenoceptors (Stjarne and Brundin, 1975). Recently, it was reported that there is a positive correlation between the human plasma levels of adrenaline and the levels of diastolic and systolic blood pressure (Franco-Morselli *et al*,

1977). These authors reported that the level of circulating adrenaline under basal conditions was significantly increased in both labile and established forms of essential hypertension.

Langer and Vogt (1971) demonstrated that under experimental conditions in which the levels of circulating adrenaline are increased by splanchnic stimulation there is a small increase in the proportion of adrenaline stored in the noradrenergic nerve endings of the peripheral nervous system. It is possible therefore that adrenaline can be stored and then subsequently be released together with the neurotransmitter noradrenaline. Recently, Rand et al (1978) reported that beta-adrenoceptor blocking agents were more effective in reducing transmitter release from the postganglionic sympathetic neuroeffector junctions when adrenaline rather than noradrenaline was stored and released from these nerve endings. It is therefore tempting to support the view that the antihypertensive effects of beta-adrenoceptor blocking agents may at least partly be due to blockade of the presynaptic beta-adrenoceptors that facilitate noradrenaline release during nerve stimulation. It is possible that in both labile and essential hypertension increased levels of circulating adrenaline (particularly under conditions of stress) enhance sympathetic nerve activity by activating presynaptic beta-adrenoceptors directly and also when adrenaline is released by nerve impulses as a co-transmitter together with noradrenaline. Under these conditions blockade of the facilitatory presynaptic beta-adrenoceptors would reduce sympathetic tone and remove the vicious circle whereby increased levels of circulating adrenaline enhance sympathetic neurotransmission.

As discussed earlier, recent reports indicate that salbutamol, a β_2-adrenoceptor agonist, has antidepressant effects in man (Widlocher et al, 1977). It is also of interest to note that beta-blocking agents like propranolol can antagonize the antidepressant effects of beta-agonists and imipramine in several animal models of depression (Puech et al, 1978). The question is still open as to whether these central effects of beta-adrenoceptor agonists are mediated through pre or postsynaptic beta-adrenoceptors. If a presynaptic facilitatory mechanism mediated by β_2 adrenoceptors exists in the central nervous system this might be a possible mechanism of action for the antidepressant effects of salbutamol. On the other hand, Puech et al (1978) suggest that the antidepressant effects of salbutamol appear to be postsynaptic because the antagonism by salbutamol of reserpine-induced hypothermia persists after inhibition of the synthesis of catecholamines by alpha-methyl-p-tyrosine.

Presynaptic Dopamine Receptors

As discussed earlier these presynaptic receptors are present in the noradrenergic nerve endings of the peripheral but not the central nervous system.

The presynaptic inhibitory dopamine receptors can be activated to reduce noradrenaline release under conditions in which the circulating levels of dopamine are increased. The latter occurs in patients under treatment with 1-dopa (Goldberg *et al*, 1973; Tyce *et al*, 1974). It is of interest that orthostatic hypotension is a common side effect in patients treated with 1-dopa (Barbeau, 1969; Calne *et al*, 1970; Goldberg, 1972) indicating some degree of impairment in the function of the sympathetic nervous system. Enero and Langer (1973) postulated that the high levels of circulating dopamine in these patients might reduce noradrenaline release in the peripheral nervous system through activation of the presynaptic dopamine inhibitory receptors on noradrenergic nerve endings. Additional support for this view was provided recently by Lokhandwala and Buckley (1978). These authors found that after the administration of 1-dopa to pentobarbital-anaesthetized dogs there was a significant inhibition of the cardiac acceleration normally elicited by electrical stimulation of cardiac sympathetic nerves. These effects of 1-dopa were prevented by prior inhibition of dopa-decarboxylase and unaffected by the inhibition of dopamine-beta-hydroxylase. In addition, the administration of the dopamine receptor blocking agent, pimozide, prevented the inhibitory action of 1-dopa on sympathetic nerve function (Lokhandwala and Buckley, 1978). Therefore it appears that presynaptic inhibitory dopamine receptors on peripheral noradrenergic nerve endings are involved in the impairment of peripheral sympathetic nerve function after treatment with 1-dopa.

References

Adler-Graschinsky, E. and Langer, S.Z. (1975). *Br. J. Pharmac.* **53**, 43–50.

Adler-Graschinsky, E. and Martinez, A.E. (1978). Actions of clonidine and propranolol on adrenergic neurotransmission in the rat cerebral cortex and in the cat superior cervical ganglion. *In* "Presynaptic Receptors" (eds. Langer, S.Z., Starke, K. and Dubocovich, M.L.), Pergamon Press, England (in press).

Arbilla, S. and Langer, S.Z. (1978a). *British J. Pharmacol.* (in press).

Arbilla, S. and Langer, S.Z. (1978b). *Nature* **271**, 559–561.

Arbilla, S. and Langer, S.Z. (1978c). *British Journal of Pharmacology* **63** (2), 389–390.

Armstrong, J.M. and Boura, A.L.A. (1973). *British J. Pharmacol.* **47**, 850–852.

Banerjee, S.P., Kung, L.S., Riggi, S.J. and Chanda, S.K. (1977). *Nature* **268**,

455–456.

Barbeau, A. (1969). *Canad. Med. Ass. J.* **101**, 791–800.

Baumann, P.A. and Maitre, L. (1977). *Naunyn-Schmiedeberg's Arch. Pharmacol.* **300**, 31–37.

Bell, C., Lang, W.J. and Laska, F. (1978). *Journal of Neurochemistry* (in press).

Braestrup, C. and Nielsen, M. (1976). *J. Pharmacol. Exp. Ther.* **198**, 596–608.

Calne, D.B., Brennan, J., Spiers, A.S.D., and Stern, G.M. (1970). *British Med. J.* **I**, 474–475.

Celuch, S.M., Dubocovich, M.L. and Langer, S.Z. (1978). *British J. Pharmac.* (in press).

Crews, F.T. and Smith, C.B. (1978). *Federation Proceedings* **37**, No. 3, 736.

Cubeddu, L.X., Barnes, E. and Weiner, N. (1975). *J. Pharmac. Exp. Ther.* **193**, 105–127.

Cubeddu, L.X. and Weiner, N. (1975). *J. Pharmac. Exp. Ther.* **192**, 1–14.

Dahlof, C., Ljung, B. and Ablad, B. (1978). *European Journal of Pharmacology* **50**, 75–78.

Delbarre, B. and Schmitt, H. (1971). *Eur. J. Pharmacol.* **13**, 356–363.

Drew, G.M., Gower, A.J. and Marriot, A.S. (1977). *Brit. J. Pharmacol.* **61**, 468.

Dubocovich, M.L. (1978). Pharmacological differences between the presynaptic alpha adrenoceptors in the peripheral and the central nervous system. *In* "Presynaptic Receptors" (eds. Langer, S.Z., Starke, K. and Dubocovich, M.L.), Pergamon Press, England (in press).

Dubocovich, M.L. and Langer, S.Z. (1974). *J. Physiol. Lond.* **237**, 505–519.

Dubocovich, M.L. and Langer, S.Z. (1976). *J. Pharmac. Exp. Ther.* **198**, 83–101.

Dubocovich, M.L., Langer, S.Z. and Pelayo, F. (1978). *Brit. J. Pharmac.* **62**, 383P.

Enero, M.A. and Langer, S.Z. (1973). *Brit. J. Pharmac.* **49**, 214–225.

Enero, M.A. and Langer, S.Z. (1975). *Naunyn-Schmiedeberg's Arch. Pharmac.* **289**, 179–203.

Farah, M.B., Adler-Graschinsky, E. and Langer, S.Z. (1977). *Naunyn-Schmiedeberg's Arch. Pharmacol.* **297**, 119–131.

Farnebo, L.O. and Hamberger, B. (1971). *Acta Physiol. Scand. Suppl.* **371**, 34–44.

Filinger, E.J., Langer, S.Z., Perec, C.J. and Stefano, F.J.E. (1978). *Naunyn-Schmiedeberg's Arch. Pharmac.* (in press).

Franco-Morselli, R., Elghozi, J.L., Joly, E., Di Giuilio, S. and Meyer, P. (1977). *Brit. Med. Journal* **2**, 1251–1254.

Goldberg, L.I. (1972). *Pharmacol. Rev.* **24**, 1–29.

Goldberg, L.I., Kohli, J.D. and Listinsky, J.J. (1978). Comparison of peripheral pre- and post-synaptic dopamine receptors. *In* "Presynaptic Receptors" (eds. Langer, S.Z., Starke, K. and Dubocovich, M.L.), Pergamon Press, England (in press).

Goldberg, L.I., Tjandramaga, T.B., Anton, A.H. and Toda, N. (1973). New investigations of the cardiovascular actions of dopamine. *In* "Frontiers in catecholamine research" (eds. Usdin, E. and Snyder, S.), Pergamon Press, New York.

Hedqvist, P. and Moawad, A. (1975). *Acta Physiol. Scand.* **95**, 494–496.

Hertting, G., Reimann, W., Zumstein, A., Jackisch, R. and Starke, K. (1978). Dopaminergic feedback regulation of dopamine release in the caudate nucleus. *In* "Presynaptic Receptors" (eds. Langer, S.Z., Starke, K. and Dubocovich, M.K.), Pergamon Press, England (in press).

Hokfelt, T., Fuxe, K., Goldstein, M. and Johansson, O. (1974). *Brain Research* **66**, 235–251.

Kobinger, W. (1978). *Rev. Physiol. Biochem. Pharmacol.* **81**, 39–100.

Langer, S.Z. (1966). *J. Pharmac. Exp. Ther.* **151**, 66–72.

Langer, S.Z. (1973). The regulation of transmitter release elicited by nerve stimulation through a presynaptic feed-back mechanism. *In* "Frontiers in Catecholamine Research (ed. Usdin, E. and Snyder, S.), pp. 543–549. Pergamon Press, New York.

Langer, S.Z. (1974a). *Medical Biology* **52**, 372–383.

Langer, S.Z. (1974b). *Biochem. Pharmac.* **23**, 1973–1800.

Langer, S.Z. (1975). Denervation supersensitivity. *Handbook of Psychopharmacology*. Vol. 2, pp. 245–279. Plenum Publishing Corporation, New York.

Langer, S.Z. (1976). *Clin. Sci. Mol. Med.* **51**, 423–426.

Langer, S.Z. (1977). *Sixth Gaddum Memorial Lecture. Br. J. Pharmac.* **60**, 481–497.

Langer, S.Z., Adler-Graschinsky, E. and Enero, M.A. (1974). Positive feed-back mechanism for the regulation of noradrenaline released by nerve stimulation. Abstract of Jerusalem Satellite Symposia. XXVI International Congress of Physiological Sciences, p. 81.

Langer, S.Z., Adler, E., Enero, M.A. and Stefano, F.J.E. (1971). *XXVth International Congress of Physiological Sciences*, p. 335, Munich.

Langer, S.Z., Adler-Grashinsky, E. and Giorgi, O. (1977). *Nature* **265**, 648–650.

Langer, S.Z. and Dubocovich, M.L. (1977). *Eur. J. Pharmac.* **41**, 87–88.

Langer, S.Z. and Dubocovich, M.L. (1978). Proceedings of the Symposium on "Peripheral Dopaminergic Receptors", Pergamon Press, Oxford (in press).

Langer, S.Z., Dubocovich, M.L. and Celuch, S.M. (1975a). Prejunctional regulatory mechanisms for noradrenaline release elicited by nerve stimulation. *In* "Chemical Tools in Catecholamine Research" (eds. Almgren, C., Carlsson, A. and Engel, J.), II, pp. 183–191. Amsterdam: Elsevier, North Holland/USA.

Langer, S.Z. and Enero, M.A. (1974). *J. Pharmac. Exp. Ther.* **191**, 431–443.

Langer, S.Z., Enero, M.A., Adler-Graschinsky, E., Dubocovich, M.L. and Celuch, S.M. (1975b). Presynaptic regulatory mechanisms for noradrenaline release by nerve stimulation. Proceeds. Symposium on Central Action of Drugs in the Regulation of Blood Pressure. (eds. Davies, D.S. and Reid, J.L.), pp. 133–151, Pitman Medical, London.

Langer, S.Z. and Luchelli-Fortis, M.A. (1977). *J. Pharmac. Exp. Ther.* **202**, 610–621.

Langer, S.Z. and Trendelenburg, U. (1966). *J. Pharmac. Exp. Ther.* **151**, 73–86.

Langer, S.Z. and Trendelenburg, U. (1969). *J. Pharmacol. Exp. Ther.* **167**, 117–142.

Langer, S.Z. and Vogt, M. (1971). *J. Physiol. Lond.* **214**, 159–171.
Langley, A.E. and Weiner, N. (1978). *J. Pharmac. Exp. Ther.* **205**, 426–437.
Ljung, B., Ablad, B., Dahlof, C., Henning, M. and Hultberg, E. (1975). *Blood Vessels* **12**, 311–315.
Lokhandwala, M.F. and Buckley, J.P. (1976). *Eur. J. Pharmac.* **40**, 183–186.
Lokhandwala, M.F. and Buckley, J.P. (1978). *J. Pharmac. Exp. Ther.* **204**, 362–370.
Long, J.P., Heintz, S., Cannon, J.G. and Kim, J. (1975). *J. Pharmac. Exp. Ther.* **192**, 336–342.
Luchelli-Fortis, M.A. and Langer, S.Z. (1975). *Naunyn-Schmiedeberg's Arch. Pharmac.* **287**, 261–275.
Medgett, I.C., McCulloch, M.W. and Rand, M.J. (1978). *Naunyn-Schmiedeberg's Arch. Pharmacol.* (in press).
Nagy, J.I., Lee, T., Seeman, P. and Fibiger, H.C. (1978). *Nature*, Vol. 274, 278–281.
Pelayo, F., Dubocovich, M.L. and Langer, S.Z. (1978). *Nature*, **274**, 76–78.
Plotsky, P.M., Wightman, R.M., Chey, W. and Adams, R.N. (1977). *Science* **197**, 904–906.
Puech, A.J., Lecrubier, Y. and Simon P. (1978). Are alpha and beta presynaptic receptors involved in mood regulation? Pharmacological and clinical data. *In* "Presynaptic Receptors" (eds. Langer, S.Z., Starke, K. and Dubocovich, M.L.), Pergamon Press, England (in press).
Rand, M.J., Majewski, H., McCulloch, M.W. and Story, D.F. (1978) An adrenaline-mediated positive feedback loop in sympathetic transmission and its possible role in hypertension. *In* "Presynaptic Receptors" (eds. Langer, S.Z., Starke, K. and Dubocovich, M.L.), Pergamon Press, England (in press).
Richards, D.A., Woodings, E.P. and Prichard, B.N.C. (1978). *Br. J. Clin. Pharmac.* **5**, 507–513.
Roach, A.G., Lefevre, F. and Cavero, I. (1978). *Clinical and Experimental Hypertension* **1(1)**, 87–101.
Roth, R.H., Walters, J.R., Murrin, L.C. and Morgenroth, V.H. (1975). Dopamine neurons: Role of impulse flow and presynaptic receptors in the regulation of tyrosine hydroxylase. *In* "Pre- and Post-synaptic Receptors" (eds. Usdin, E. and Bunney, W.E.), pp. 5–46. Marcel Dekker, New York.
Saavedra, J.M., Palkovits, M. and Brownstein, M.J. (1974). *Nature* **248**, 695–696.
Scatton, B., Pelayo, F., Dubocovich, M.L., Langer, S.Z. and Bartholini, G. (1978). *In* "Presynaptic Receptors" (eds. Langer, S.Z., Starke, K. and Dubocovich, M.L.), Pergamon Press, England (in press).
Schildkraut, J.J., Kety, S.S. (1967). *Science* **156**, 21–30.
Starke, K. (1977). *Rev. Physiol. Biochem. Pharmac.* **77**, 1–124.
Starke, K., Montel, H., Gay, K.W. and Merker, R. (1974). *Naunyn-Schmiedeberg's Arch. Pharmac.* **285**, 133–150.
Stjarne, L. and Brundin, J. (1975). *Acta Physiol. Scand.* **94**, 139–141.
Stjarne, L. and Brundin, J. (1976a). *Acta Physiol. Scand.* **97**, 88–93.
Stjarne, L. and Brundin, J. (1976b). *Acta Physiol. Scand.* **97**, 267–269.

Stjarne, L. and Brundin, J. (1977). *Acta Physiol. Scand.* **101**, 199–210.

Stjarne, L. and Gripe, K. (1973). *Naunyn-Schmiedeberg's Arch. Pharmac.* **280**, 441–446.

Story, D.F., Briley, M.S. and Langer, S.Z. (1978). *In* "Presynaptic Receptors" (eds. Langer, S.Z., Starke, K. and Dubocovich, M.L.), Pergamon Press, England (in press).

Sulser, F., Vetulani, J. and Mobley, P.L. (1978). *Biochemical Pharmacology*, **27**, 257–261, Pergamon Press.

Taube, H.D., Starke, K. and Borowski, E. (1977). *Naunyn-Schmiedeberg's Arch. Pharmacol.* **299**, 123–141.

Trendelenburg, U. (1963). *Pharmacol. Rev.* **15**, 225–276.

Tyce, G.M., Sharpless, N.S. and Muenter, M.D. (1974). *Clin. Pharmacol. Ther.* **16**, 782–788.

Vetulani, J., Stawarz, R.J., Dingell, J.V. and Sulser, F. (1976). *Naunyn-Schmiedeberg's Arch. Pharmacol.* **293**, 109–114.

Vogel, S.A., Silberstein, S.D., Berve, K.R. and Kopin, I.J. (1972). *Eur. J. Pharmac.* **20**, 308–311.

Von Euler, U.S. (1946). *Acta Physiol. Scand.* **12**, 73–97.

Weinstock, M., Thoa, N.B. and Kopin, I.J. (1978). *Eur. J. Pharmac.* **47**, 297–302.

Widlocher, D., Lecrubier, Y., Jouvent, R., Puech, A.J. and Simon, P. (1977). *The Lancet*, 8 October, 767–768.

Yamaguchi, N., De Champlain, J. and Nadeau, R.A. (1977). *Circulation Research* **41**, 108–117.

HYPERTENSION, HYPOTENSION AND THE AUTONOMIC NERVOUS SYSTEM

P.S. SEVER

Medical Unit, St. Mary's Hospital, London, U.K.

Introduction

Blood pressure in man results from a complex interplay of physiological and endocrine functions which control cardiac output and peripheral vascular resistance. When the control of arterial pressure fails, as in hypertension on the one hand, or autonomic failure at the other extreme, it is changes in the peripheral vascular resistance which constitute the major functional abnormality. The sympathetic nervous system is a major determinant of vascular tone, and the integrity of this system and its activity in health and disease have for many years been the province of the physiologist with his elaborate tests of the various reflex pathways involving this system.

Naturally, several attempts have been made to determine the amounts of the mediator of sympathetic effector responses as an alternative method for evaluating sympathetic activity. However, progress in this field has been frustrated by the lack of suitable methods to determine the minute quantities of the neurotransmitter, noradrenaline, that are present in human plasma.

The early methods lacked sensitivity and specificity. Bioassay, although sensitive, suffered from the lack of absolute specificity for individual catecholamines, and the early chemical methods which were colorimetric lacked both sensitivity and specificity. The development of fluorimetric assays was a great advance on previous methods. However, at the lower concentrations of noradrenaline, such as those found in human plasma, errors were magnified and the reproducibility of assays was a major problem.

Engelman et al (1968) published details of an elaborate double-isotope-derivative radio-enzymatic assay for catecholamines, which has since been modified to permit the isolation and quantitation of individual catecholamines with a remarkable degree of specificity and sensitivity (Engelman and Portnoy, 1970). Henry and co-workers (1975) reported details of another radio-enzymatic procedure which was specific for noradrenaline, and it is a modification of this method which is currently in use in our laboratory.

Methodology notwithstanding, several important factors have to be considered before plasma catecholamine measurements can be considered as true reflections of the activity of the sympathetic nervous system.

There is evidence that because of the blood-brain barrier noradrenaline synthesised in the brain does not reach the plasma in significant amounts (Axelrod 1971). The adrenal glands could contribute to the quantum of noradrenaline in plasma; however, adrenaline is the dominant amine released from adrenals, and bilateral adrenalectomy produces no apparent decline in plasma noradrenaline concentration (Roizen et al 1974).

A complex series of events surrounds the release of noradrenaline from sympathetic nerve endings.(Iversen 1967). Most of the neuro-transmitter released on activation of sympathetic efferent pathways

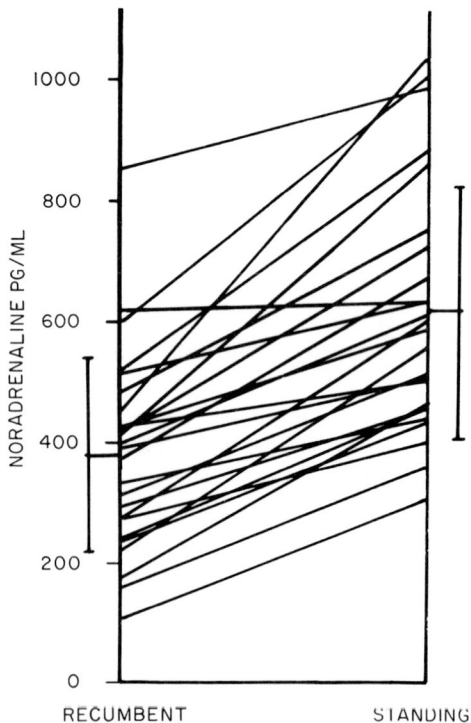

Fig. 1 Plasma noradrenaline concentration determined in 25 normal subjects following one hour's recumbency and then five minutes' standing. Means values ± SD are given.

is taken up into the nerve ending and re-stored. Extra-neuronal uptake and metabolism by o-methylation, oxidative deamination and conjugation are also important routes by which noradrenaline is inactivated (Iversen 1967). Only a small fraction of the noradrenaline initially released from nerve endings escapes from the synaptic cleft, and it is this fraction which may enter the circulation.

Despite these theoretical objections to the validity of plasma catecholamine measurements as indicators of sympathetic tone, studies in normal individuals have demonstrated that conventional stimuli to the sympathetic nervous system evoke physiological responses that are paralleled by changes in the circulating concentration of noradrenaline.

Activation of the baro-receptor reflexes by a change in posture is associated with an increase in the circulating plasma noradrenaline concentration (Fig. 1) (Osikowska and Sever 1976). This 'switching on' of the sympathetic system by postural change can be shown to be a gradual response. In a recent study (Rosenthal et al 1977) the changes in plasma noradrenaline concentration were observed during a stepwise tilt through an angle of 45°. It can be seen from this study

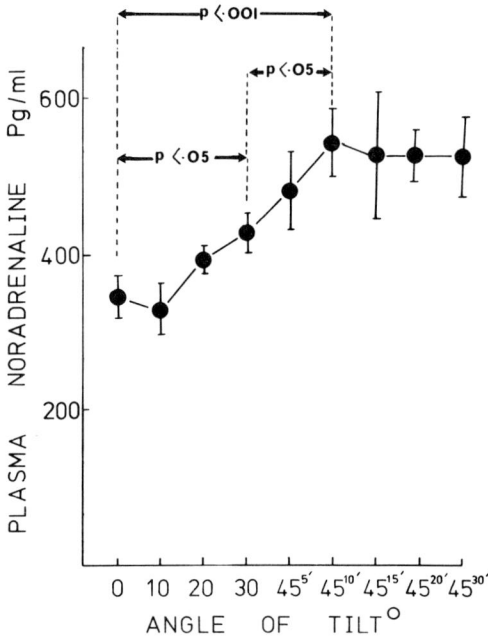

Fig. 2 Plasma noradrenaline during gradual tilt in 10 subjects. Values given represent mean ± SEM.

(Fig. 2) that there are incremental rises as the angle of tilt is increased. After a 45° tilt for a period of 10 minutes the raised levels are sustained for between 30 and 60 minutes. In this particular study the changes in plasma noradrenaline paralleled the increase in mean arterial pressure (Fig. 3).

It has been shown by Lake et al (1976) and in similar studies carried out in our own laboratory that isometric exercise (Vecht et al 1977) (Fig. 4) and exposure to cold (the cold-pressor response) are also associated with increases in plasma noradrenaline concentration.

Hypotension

Orthostatic hypotension is a disabling feature of degeneration of the autonomic nervous system, which may occur as an isolated disturbance (idiopathic orthostatic hypotension) or in association with other neurological problems (Shy-Drager syndrome or multiple system atrophy). Orthostatic hypotension is also a common side effect of treatment with sympatholytic drugs acting at a ganglionic site (ganglion blocking drugs) or on the postganglionic adrenergic neurone (adrenergic neurone blocking drugs).

Whether the integrity of the efferent sympathetic pathway is

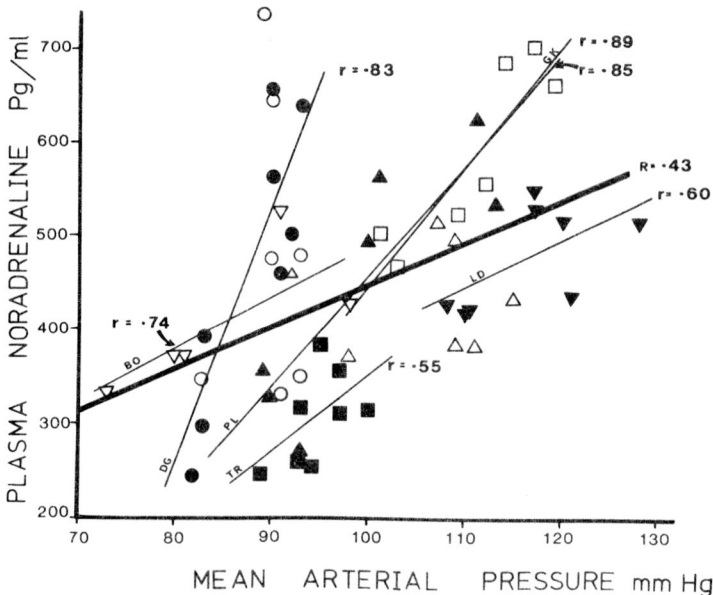

Fig. 3 Relationship between plasma noradrenaline and mean arterial pressure during gradual tilt in 8 subjects. For all subjects R = 0.43 (p < 0.001).

compromised by a pathological process or by pharmacological agents the result is the interruption of the reflex-mediated increase in the firing rate of the sympathetic neurones to produce vasoconstriction when the erect posture is assumed. We have already demonstrated that in the normal individual this is associated with a rise in noradrenaline in plasma, and it is not surprising that in autonomic disease this rise is abolished. In Fig. 5 the response to a 60° tilt has been monitored in 10 patients with autonomic disease (Bannister et al 1977). Not only was there a failure in noradrenaline increase with tilt, but basal levels of noradrenaline in plasma were significantly lower than in age-matched control subjects. The average reduction of resting plasma noradrenaline to 20% of normal corresponds approximately to the reduction to 20% of the preganglionic neurons of the intermediolateral columns of the spinal cord found at post mortem in patients with multiple system atrophy.

Comparable results have been recorded in other groups of patients with defective autonomic pathways, such as tetraplegics (Mathias et al 1975) and familial dysautonomia (Zeigler et al 1977).

It is not surprising that when postural hypotension has been produced by drugs such as guanethidine and bethanidine similar

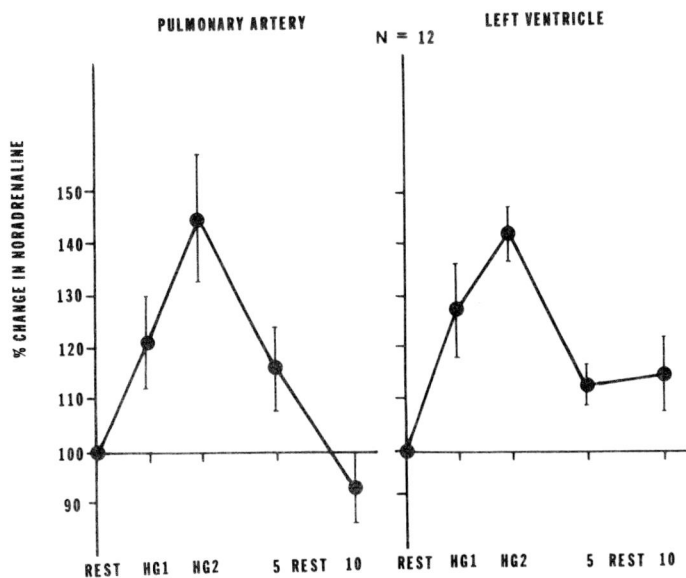

Fig. 4 Percentage changes in plasma noradrenaline ($\pm$ SEM) in response to isometric exercise. Samples taken in 12 subjects from pulmonary artery and left ventricle at cardiac catheterisation. HG 1 and HG 2 represent 1½ minutes and 3 minutes of handgrip (30% maximal grip). Further samples during 5 and 10 minutes' rest.

reductions in plasma noradrenaline can be demonstrated with abolition of the postural response (P. Sever, unpublished data).

Plasma noradrenaline estimation may therefore represent a useful pointer to the diagnosis and progression of autonomic disease.

From these observations and other experimental data it would appear that basal noradrenaline concentration and the changes induced by posture and other stresses are useful determinants of sympathetic tone.

Hypertension

Overactivity of the sympathetic nervous system has been implicated in the pathogenesis of essential hypertension. Provocative manoeuvres, in particular the use of postural challenge, have suggested inappropriate activation of the sympathetic system in some subjects in the early stages of essential hypertension (Frohlich et al 1967; Esler and Nestel 1973). There is also more indirect evidence for a contribution of the sympathetic system to essential hypertension in that the responses of these subjects to anti-hypertensive, antiadrenergic drugs are enhanced (Doyle and Smirk 1955).

The application of plasma catecholamine estimations to the investigation of patients with this condition has produced conflicting

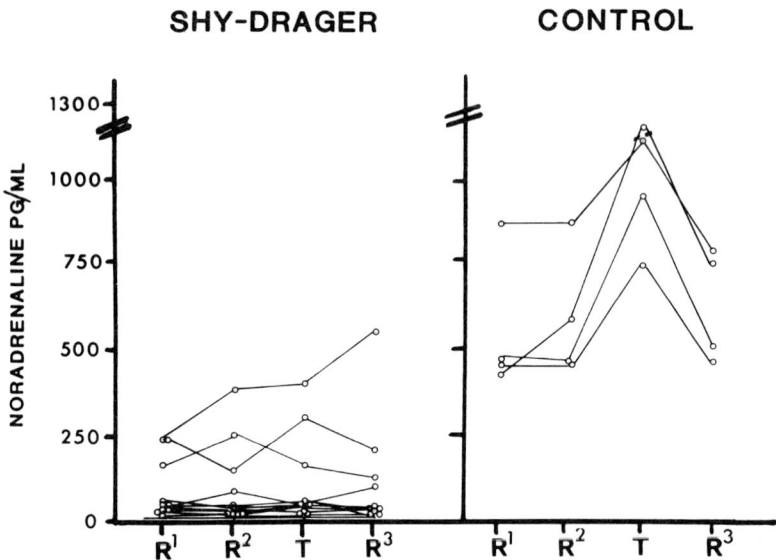

Fig. 5   Plasma noradrenaline determined in 10 patients with Shy-Drager syndrome and 4 control subjects (age matched). R1 and R2 — 1 and 2 hours' recumbency; T — 10 minute 60° tilt; R3 — 1 hours' recumbency.

results. The earlier reports of Engelman *et al* (1970), De Quattro and Chan (1972), Louis *et al* (1973) and de Champlain *et al* (1976) claimed abnormally high levels of plasma noradrenaline in patients with essential hypertension.

Overinterpretation of these early findings was questioned, in view of the absence of suitable control populations. Constitutional factors, including age, sex and race, had not been taken into consideration in the selection of control subjects in many of these studies, and the use of laboratory personnel as control subjects is far from ideal.

Zeigler *et al* (1976) reported that plasma noradrenaline concentration varied with age, and a similar study from our unit has confirmed this observation (Sever *et al* 1977). In fact, the failure of previous authors to age-match control subjects and hypertensives could have wholly accounted for the observed differences in catecholamine levels in these two groups.

In a preliminary study (Sever *et al* 1977) untreated patients with essential hypertension were investigated, and these subjects were compared with age-matched control normotensive subjects who were randomly selected from a healthy Civil Service population. All investigations were performed between the hours of 9 and 10.30 in the morning in an attempt to minimise any possible changes due to diurnal rhythms, and all subjects, both hypertensive and normotensive, were investigated under identical conditions − initially following a period of recumbency and subsequently after a 5 minute period of standing.

A linear relationship was found between age and plasma noradrenaline in normotensives during recumbency and in the upright posture (Fig. 6a, 7a). The significance of this correlation was lost in hypertensives, and this was due to raised plasma noradrenaline levels in many younger patients.

The data from this initial study have now been extended, and analysis of investigations carried out on 70 hypertensive subjects has revealed major racial differences in plasma catecholamine disposition in essential hypertensives (Fig. 6 and 7). The normal positive correlation between age and plasma noradrenaline is seen in Negro hypertensive subjects, but in Caucasian hypertensives age does not appear to be a determinant of plasma noradrenaline.

On the basis of the available evidence we might infer that in younger white subjects with essential hypertension the abnormally raised plasma noradrenaline levels could reflect an increase in sympathetic tone. Whether or not this observation is consistent with an hypothesis that sympathetic overactivity plays an aetiological

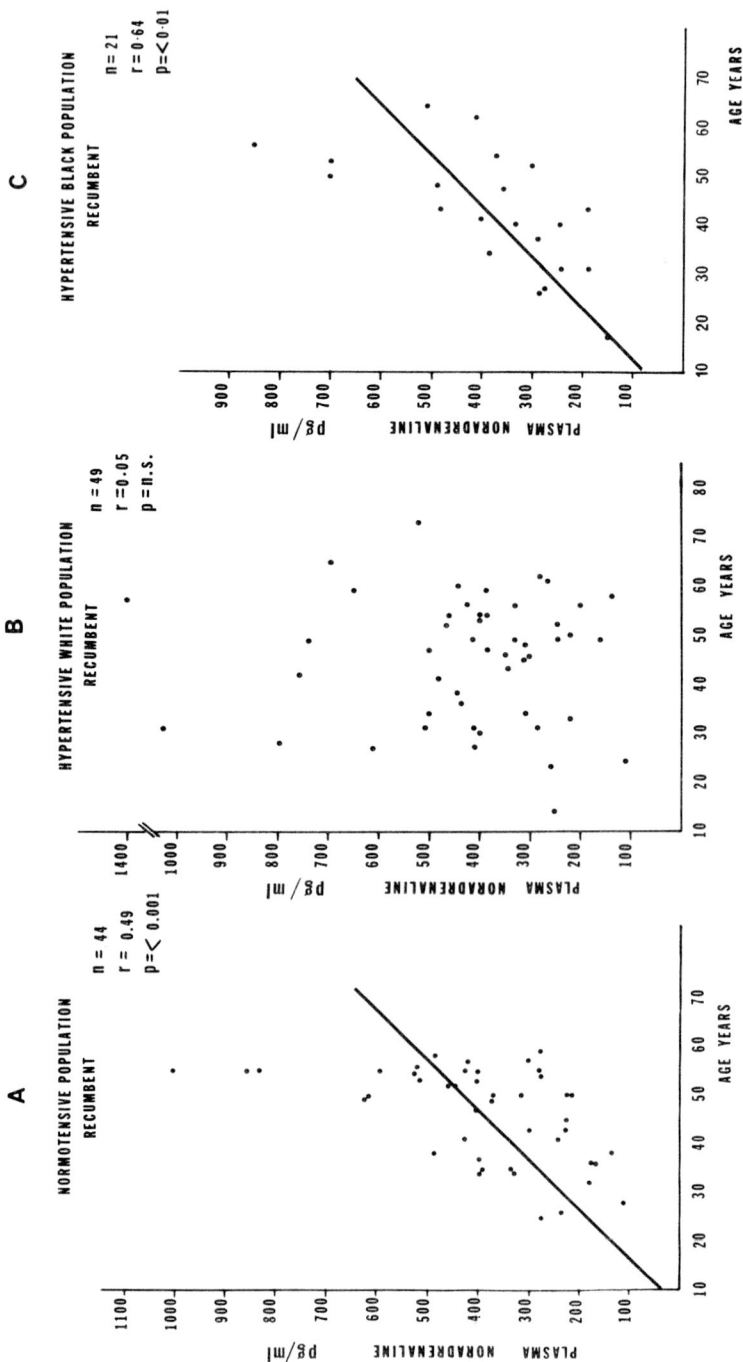

Fig. 6 Plasma noradrenaline and age in normotensive (white) and hypertensive (black and white) populations. Levels determined following 60 minutes' recumbency.

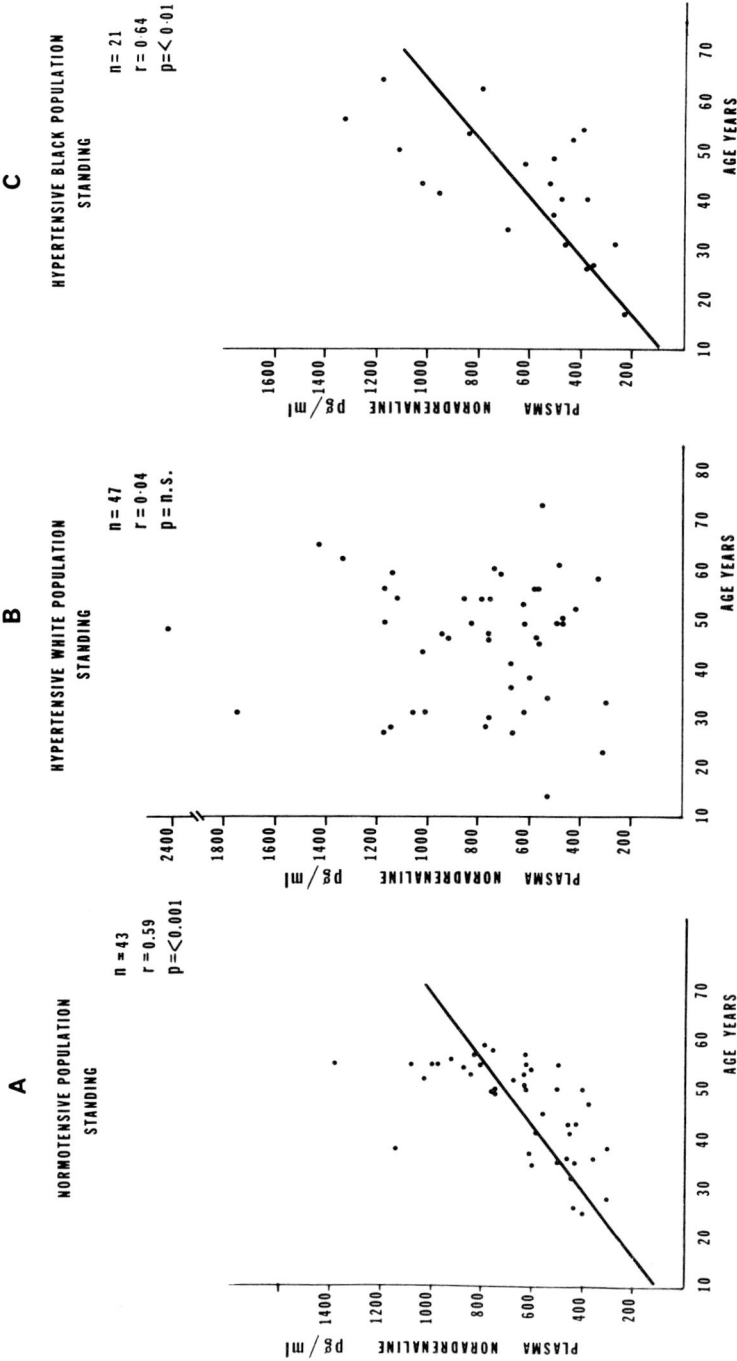

Fig. 7 Plasma noradrenaline and age in normotensive (white) and hypertensive (black and white) populations. Levels determined after 5 minutes' standing.

role in the development of essential hypertension remains to be clarified by further studies. For the reasons outlined above, alterations in catecholamine metabolism, neuronal or extra-neuronal uptake, could well account for the variations with age in the normotensive subjects; nevertheless, it is tempting to conclude that the abnormalities in the hypertensive subjects were indeed a reflection of increased sympathetic nerve activity.

Summary

In this brief review I have attempted to demonstrate how the advent of sensitive and specific catecholamine estimations may contribute to the investigation of disorders of sympathetic function. Nevertheless, it must be emphasised that abnormalities in plasma catecholamine disposition do not provide any insight into the level at which the sympathetic nervous system may be involved. Progress in this field requires utilisation of these biochemical methods in conjunction with the refined neurophysiological techniques now available for recording sympathetic neuronal activity.

References

Axelrod, J. (1971). *Science* **173**, 598–606.
Bannister, R., Sever, P.S. and Gross, M. (1977). *Brain* **100**, 327–344.
de Champlain, J., Farley, L., Cousineau, D. and Ameringen, M-R. (1976). *Circulations Res.*, 109–114.
De Quattro, V. and Chan, S. (1972). *Lancet* **1**, 806–809.
Doyle, A.E. and Smirk, F.H. (1955). *Circulation* **12**, 543–622.
Engelman, K., Portnoy, B. and Lovenberg, W. (1968). *Am. J. Med. Sci.* **255**, 259–268.
Engelman, K. and Portnoy, B. (1970). *Circulation Res.* **26**, 53–57.
Engelman, K., Portnoy, B. and Sjoerdsma, A. (1970). *Circulation Res.* **27** (Suppl. 1), 141–146.
Esler, M.D. and Nestel, P.J. (1973). *Clin. Sci.* **44**, 213–226.
Frohlich, E.D., Tarazi, R.C., Ulrych, M., Dustan, H.P. and Page, I.H. (1967). *Circulation* **36**, 387–393.
Henry, D.P., Starman, B.J., Johnson, D.G. and Williams, R.H. (1975). *Life Sci.* **16**, 375–384.
Iversen, L.L. (1967). The uptake and storage of noradrenaline in sympathetic nerves. Cambridge University Press.
Lake, C.R., Zeigler, M.G. and Kopin, I.J. (1976). *Life Sci.* **18**, 1315–1326.
Louis, W.J., Doyle, A.E. and Anavekar, S. (1973). *New Engl. J. Med.* **288**, 599–601.
Mathias, C.J., Christensen, N.J., Corbett, J.L., Frank, H.L., Goodwin, T.J. and Peart, W.S. (1975). *Clin. Sci. Mol. Med.* **49**, 291–299.
Osikowska, B. and Sever, P.S. (1976). *Brit. J. Clin. Pharmacol.* **3**, 963–964.

Roizen, M.F., Moss, J., Henry, D.P. and Kopin, I.J. (1974). *Anesthesiology* **41**, 432–439.

Rosenthal, T., Birch, M., Osikowska, B. and Sever, P.S. (1977). *Cardiovascular Res.* (in press).

Sever, P.S., Osikowska, B., Birch, M. and Tunbridge, R.D.G. (1977). *Lancet* **1**, 1077–1081.

Vecht, R.J., Graham, G.W.S. and Sever, P.S. (1977). *Brit. Heart J.* (in press).

Zeigler, M.G., Lake, C.R. and Kopin, I.J. (1976). *Nature* **261**, 333–335.

Zeigler, M.C., Lake, C.R. and Kopin, I.J. (1976). *New Engl. J. Med.* **294**, 630–633.

THE DIAGNOSIS AND MANAGEMENT OF CHRONIC AUTONOMIC FAILURE

SIR ROGER BANNISTER

The National Hospital, Queen Square, London, U.K.

Introduction

In the past 10 years there have been considerable advances in our under-standing of the autonomic nervous system. In this chapter I shall con-centrate on studies of chronic autonomic failure, but in order to put them into perspective I have listed some of the important syndromes in Table I. The chronic progressive disorders form a distinct group and include 'neurogenic' or idiopathic orthostatic hypotension, des-cribed first by Bradbury and Eggleston (1925); autonomic failure with multiple system atrophy, identified by Shy and Drager (1960); and most recently autonomic failure with parkinsonism, the first case described by Fichefet *et al* (1965). Selective involvement of autonomic fibres has long been known to occur in diabetes, alcoholism and neurosyphilis and in certain general diseases such as amyloid. More recently it has been realised that there are also acute and sub-acute sympathetic and parasympathetic neuropathies (Tomashefsky *et al* 1972; Hopkins *et al* 1972), and others induced by drugs such as vincristine (McLeod and Penny, 1969, and acrylamide (Fullerton and Barnes, 1966). As techniques of investigation become more refined more syndromes are likely to be identified.

TABLE I

Syndromes of Autonomic Failure with Postural Hypotension

1.	Multiple system atrophy (Shy-Drager Syndrome).
2.	'Parkinsonian' autonomic failure.
3.	'Neurogenic' orthostatic hypotension.
4.	Acute and subacute, sympathetic and parasympathetic peripheral neuropathies.
5.	Diabetes.
6.	Neurosyphilis
7.	Amyloid.

The syndromes of chronic autonomic failure, though rare, are more common than is generally recognised and are difficult to manage. Moreover they throw some light on the curious neuronal degenerations which make up so much of the field of neurology. As Claude Bernard commented 'numberless pathological lesions are real experiments'. The problems of localisation of the lesion in these conditions illustrate the usefulness of the study of disease to unravel more about normal structure and function in the autonomic nervous system.

Clinical Features and Investigation

The clinical presentation of the syndromes of chronic autonomic failure is usually with attacks of loss of consciousness on standing or walking, or a form of parkinsonism, although the tremor is often atypical, with added cerebellar features. Realisation that the patients in the latter group do not have true Parkinson's disease often comes when they develop symptoms suggestive of cerebral ischaemia on standing, sometimes precipitated by l-Dopa treatment, or complain of impotence, disturbances of bladder function or defective sweating. In a patient who presents such features one of the important special investigations to confirm the diagnosis is monitoring of blood pressure with an arterial catheter, because sufficiently precise arterial pressure recording is impossible by conventional sphygmomanometry.

Figure 1 shows a tracing from such a patient with a brachial artery catheter. His blood pressure could be maintained when he was sitting or standing, provided an inflated antigravity suit was used to protect him against pooling of blood in the extremities. Without this support his blood pressure fell continuously and he would eventually lose consciousness.

Figure 2 shows a simplified pictorial representation of some of the cardiovascular reflexes (from Gray, 1972) indicating the baroreceptor pathways, the vasomotor 'centre' in the brain stem, and the efferent pathways, those in the vagus to slow the heart rate, and the sympathetic efferents to increase heart rate and to cause vasoconstriction of vascular smooth muscle. A battery of tests has been designed to investigate the cardiovascular reflexes (see Johnson & Spalding, 1974); the most important are the responses of blood pressure to tilting and the Valsalva manoeuvre.

Table II lists three types of lesion in patients with postural hypotension; the most important is that of the sympathetic efferent fibres to capacity and resistance vessels in muscles and the splanchnic area. A lumbar sympathectomy has little effect on blood pressure, and

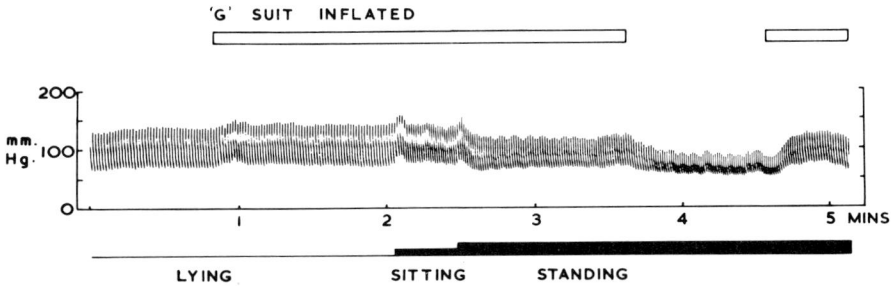

Fig. 1 The effect of an antigravity suit on the changes of arterial blood pressure which occur on sitting and standing in a patient with chronic autonomic failure. (From Bannister *et al*, 1969, used with permission).

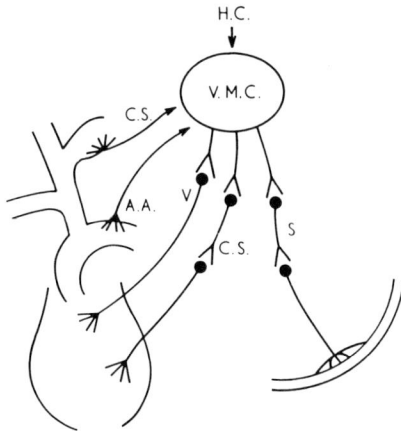

Fig. 2 Diagram of neural control of blood pressure. (From Gray, I.R. (1972) used with permission). C.S. = carotid sinus; A.A. = aortic arch; V.M.C. = vasomotor centre; H.C. = higher centres; V = vagus; C.S. = cardiac sympathetic; S = sympathetic nerves.

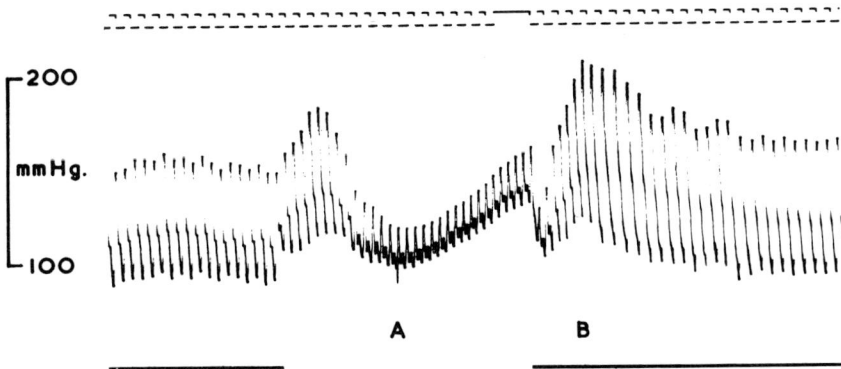

Fig. 3 Valsalva manoeuvre in a patient with normal baroreceptor reflexes, showing a) halting of fall of pulse pressure with tachycardia in phase II; b) 'overshoot' of systolic and diastolic pressures and bradycardia in phase IV. Upper trace – time in seconds; middle trace – arterial pressure; lower trace – time marker for Valsalva manoeuvre. (From Bannister *et al* 1977, used with permission).

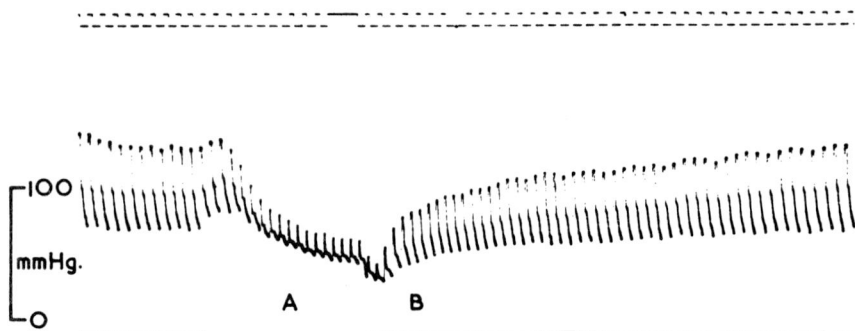

Fig. 4 Valsalva manoeuvre showing very abnormal response in chronic autonomic failure: a) continuing fall of pulse pressure throughout phase II; b) 'blocked' response with slow mechanical recovery in phase IV without evidence of 'overshoot' or bradycardia. (From Bannister *et al* 1977, used with permission).

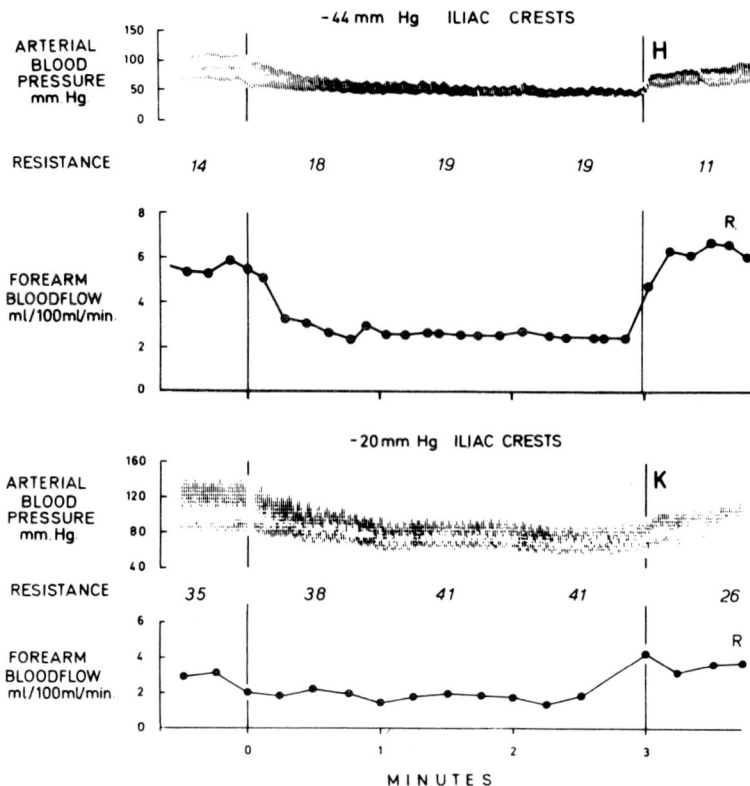

Fig. 5 The effect of suction on arterial blood pressure, forearm resistance and forearm blood flow in a normal subject (upper part), and in a patient with chronic autonomic failure (lower part). (From Bannister *et al* 1967, used with permission).

TABLE II

Relative Contribution of Autonomic Lesions to Postural Hypotension

1. Lesion of sympathetic vasoconstrictor efferent fibres to capacity and resistance vessels of limbs and splanchnic areas.
2. Lesion of sympathetic cardio-accelerator fibres.
3. Lesion of parasympathetic efferent fibres.

TABLE III

Tests of Autonomic Function

1. Sympathetic efferent constrictor fibres to capacity and resistance vessels
   a) postural hypotension
   b) lack of overshoot on Valsalva test
   c) lack of blood pressure rise on stress
   d) low resting plasma noradrenaline
   e) lack of rise of plasma noradrenaline on tilting

2. Sympathetic efferent fibres to heart
   a) lack of tachycardia during Valsalva, phase II
   b) lack of tachycardia on tilting
   c) lack of tachycardia on stress or exercise

3. Parasympathetic efferent fibres to heart
   a) lack of sinus arrhythmia
   b) lack of effect of carotid massage
   c) lack of rise of cardiac rate with atropine

even a combined lumbar and cervical sympathectomy has only a transient effect. The splanchnic area is of critical importance and loss of the major part of the total sympathetic outflow is probably necessary before postural hypotension occurs (Low *et al* 1975).

Table III shows some tests for lesions of these groups of fibres. First, with a lesion of the sympathetic efferent fibres, postural hypotension is present and no overshoot occurs in phase IV of the Valsalva, because there is no vasoconstriction of peripheral circulation into which the increased cardiac output is pumped (Fig. 3 and 4). There is also no rise of blood pressure on stress and also a low resting plasma noradrenaline level without any increase on tilting. Table III also shows some tests for the sympathetic efferent fibres to the heart. There has been some uncertainty as to the cause of the tachycardia during the Valsalva manoeuvre, but it is

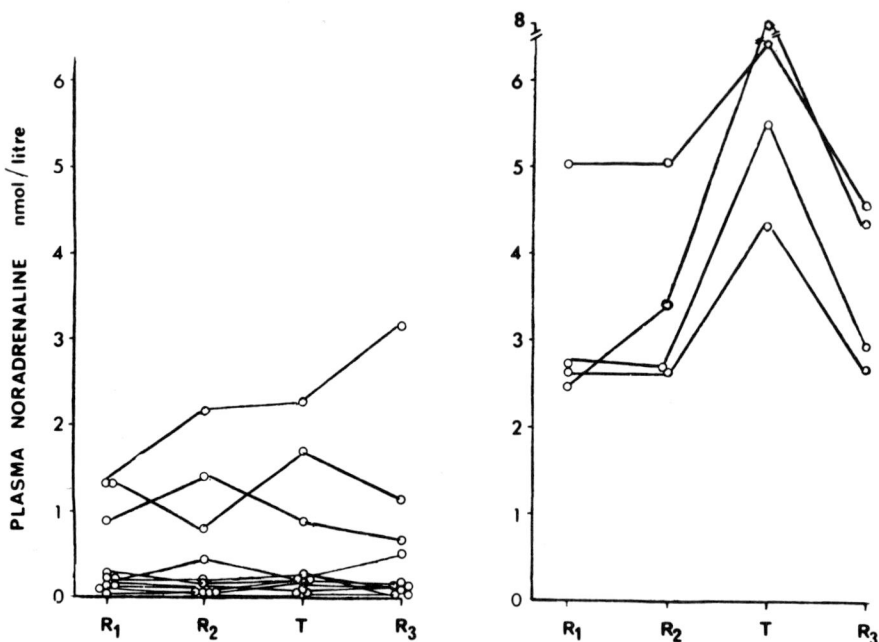

Fig. 6   The effect of tilt on plasma noradrenaline (nmol/L) in patients with autonomic failure and control subjects. R_1 and R_2 denote recordings following one and two hours' recumbency respectively, T follows the ten-minute $60°$ tilt, and R_3 a further one hour's recumbency. (From Bannister *et al* 1977, used with permission.)

blocked by propranolol and unaffected by atropine, and so is probably principally sympathetic. The tachycardia on tilting and on stress or exercise is probably mediated by similar efferent pathways. Probably least important in relation to postural hypotension are the parasympathetic vagal efferents to the heart, necessary for normal sinus arrhythmia and for the bradycardia caused by carotid sinus massage.

All these tests are listed as though they were purely for efferent pathways, but most involve reflexes and hence have central and afferent connections as well. The next stage, if postural hypotension and a blocked Valsalva manoeuvre are demonstrated, is to try to show whether the lesion is afferent or efferent (Table IV). If the vaso-constrictor response to stress is preserved the efferent limb is intact and the lesion is probably afferent or central. If the sweating response to generalised body heating is preserved it can be argued that the lesion is probably afferent since sweat glands are innervated by sympathetic (cholinergic) fibres. The loss of sweating and pilo-

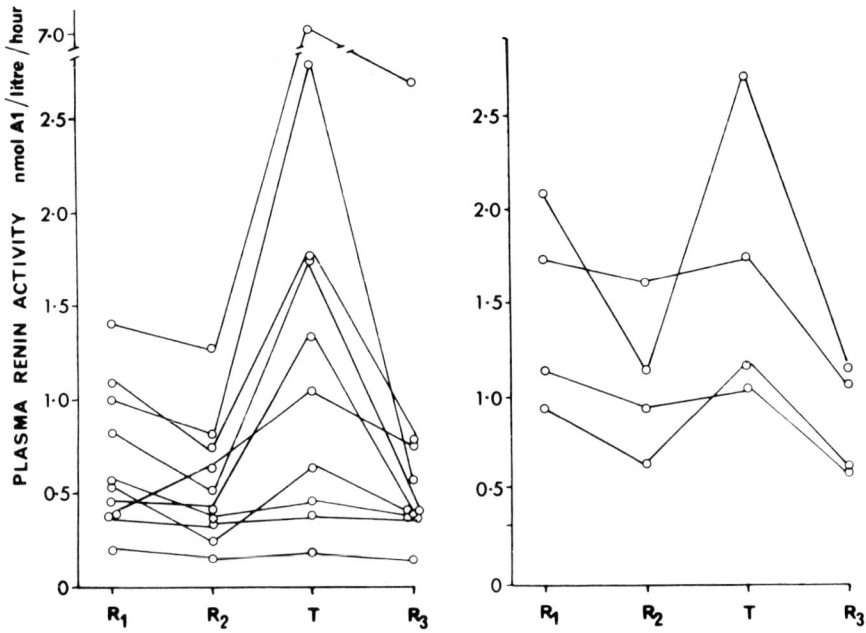

Fig. 7 The effect of tilt on plasma renin activity (nmol Angiotensin I/1/h) in patients with autonomic failure and control subjects. R_1 and R_2 denote recordings following one and two hours' recumbency respectively, T follows the ten minute $60°$ tilt, and R_3 a further one hour's recumbency. (From Bannister *et al* 1977, used with permission.)

erection after intradermal acetylcholine injections has been taken as evidence of a post-ganglionic lesion, but these tests are not easy to interpret (Barany and Cooper 1956). There may be a single lesion on a single pathway in autonomic failure, but lesions usually occur at more than one site and may be partial rather than complete. At present it is not usually possible in the presence of an efferent lesion to diagnose with certainty the presence of another lesion on the afferent side of the reflex arc, but this can sometimes be achieved (Bannister *et al* 1967, Aminoff *et al* 1971).

Identification of precisely which fibres are defective is more difficult and involves techniques which are not easy to use in the routine study of patients. The lower body suction technique has been used to simulate the effects of standing (Bannister *et al* 1967). The patient then remains in the recumbent position with an arterial catheter in place, and forearm blood flow is measured by plethysmography. Suction mimics the shift of blood which occurs

TABLE IV

Principles of Localisation of Lesions in Autonomic Failure

	Observation	*Conclusion*
1.	Vasoconstrictor response to stress	afferent lesion
2.	Defective thermal sweating	efferent lesion
3.	Normal pilo-erection to intradermal acetylcholine	preganglionic lesion
4.	Defective pilo-erection to intradermal acetylcholine	postganglionic lesion

on standing. Figure 5 (upper tracing) shows that if negative pressure is applied in a normal subject, to the level of the iliac crest, there is a six fold increase in the resistance in the forearm muscle blood vessels. Figure 5 (lower tracing) shows a patient with autonomic failure in whom forearm blood flow hardly changes in response to negative pressure applied to the lower half of the body. There is a slight fall but this is a reflection of the drop in perfusion pressure.

Figure 6 shows the plasma noradrenaline levels in 10 patients with progressive autonomic failure in association with multiple system atrophy or parkinsonism. The results in the patients are compared with those in 4 control subjects. In all the patients with autonomic failure the level is lower than in the controls and in 7 out of the 10 it

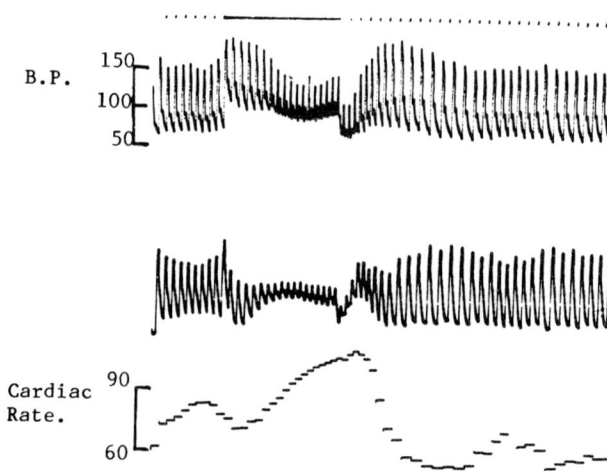

Fig. 8 Valsalva response in normal subject showing blood pressure (upper trace), photoplethysmograph from finger (middle trace), cardiac beat-to-beat record (lower trace).

is very close to the lowest threshold of measurement. There are numerous problems in determining rates of noradrenaline release from levels in the plasma (see Sever, this volume, p. 53), but a reduction in the plasma level is likely to be caused by defective release from sympathetic efferent terminals in blood vessels and not as a result of changes in uptake elsewhere or diffusion across the blood brain barrier or from any other site. The patients in whom there was a measurable level of noradrenaline, and in one case some response to tilt, were patients who had, according to the other criteria, a less complete sympathetic efferent paralysis. Renin stores persist in chronic autonomic failure (Wilcox *et al* 1974) and figure 7 shows the plasma renin activity in 10 patients and the response to tilt for 10 minutes. In some patients the resting level is low and unchanged on tilt but in those in whom the resting plasma renin is high there tends to be a marked rise, higher indeed than in some control subjects. If there can be assumed to be only a few functioning sympathetic efferent fibres to the kidney then renin from stores may be released by auto-regulatory renal vasodilation when the perfusion pressure falls, which has also been suggested by Mathias *et al* (1977).

The second reflex system to be considered in autonomic failure is that of the sympathetic rate control of the heart. Figures 8 and 9 show Valsalva responses with a radial artery pressure recording using a 21 gauge cannula, which is more stable than brachial artery catheterisation with a needle. In addition there is a finger plethysmographic recording and a cardiac beat-to-beat recording. The Valsalva manoeuvre (Fig. 9) from a patient with autonomic failure shows neither a normal rise in pulse rate in phase II (sympathetic) nor a fall in phase IV (parasympathetic).

The third relevant reflex system in autonomic failure is the para-sympathetic rate control of the heart. Figure 10 shows a cardiac beat to beat recording during deep breathing. In a normal subject sinus arrhythmia may be as great as 20—30 beats with deep breathing and is measurable during shallow breathing. This tests the vagal parasympathetic supply to the heart (Wheeler and Watkins 1972) and is absent in chronic autonomic failure.

From physiological studies in cases of progressive autonomic failure with multiple system atrophy or parkinsonism there is good evidence of an efferent sympathetic lesion of varying severity and some evidence that this is both pre-ganglionic and post-ganglionic, though more precise methods are needed to study the post-ganglionic pathway. The sympathetic efferent lesion affects resistance and capacity vessels and the heart, and is usually more severe than

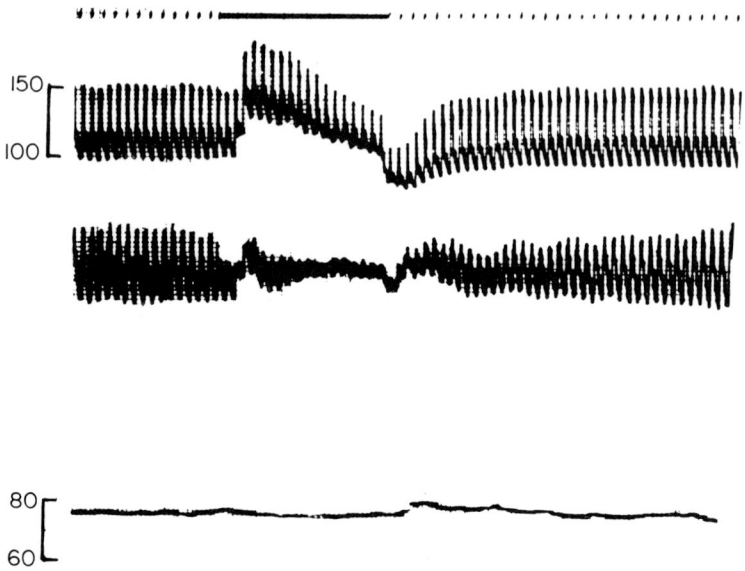

Fig. 9   Valsalva response in patient with chronic autonomic failure. Records as in Fig. 8.

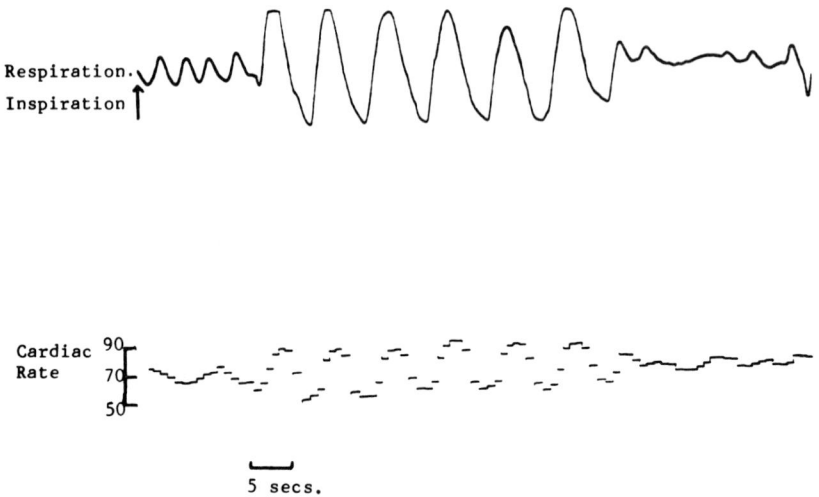

Fig. 10   Sinus arrhythmia in normal subject. Respiratory record (upper trace) and cardiac beat-to-beat (lower trace).

the vagal central or efferent lesion. This makes an interesting and important contrast to diabetic autonomic neuropathy, in which the vagal efferent lesion is usually more severe than the sympathetic efferent lesion.

Treatment

Treatment of patients with progressive autonomic failure is fraught with difficulties. Their parkinsonian features do not usually respond to 1-Dopa (which may aggravate the postural hypotension) or to anticholinergic drugs, presumably because the central lesion differs from that of idiopathic parkinsonism. Nevertheless the usual remedies for parkinsonism, including Sinemet, must be tried in each patient and in some there is an improvement. Equally difficult is management of their orthostatic hypotension (Bannister *et al* 1969). There are several methods of combating this, of varying degrees of subtlety. Figure 11 shows the mean blood pressure before and after lower body suction of 4 patients and normal subjects with forearm resistance, calculated, before and after suction. Different treatment conditions were studied; first the inflatable anti-gravity suit (Fig. 12) which reduces the vascular volume for pooling of blood in the lower part of the body. This treatment reduced the blood pressure fall during suction, without any change in the contractility of the forearm vessels. However this has the disadvantage of reducing the passive resistance or myogenic response to stretch which is intrinsic even to denervated vessels (Folkow and Neol 1971) and it may remove the effect of 'practice' in vasoconstrictor responses of blood vessels, lack of which is seen in healthy subjects after enforced recumbency of several weeks. Cerebral ischaemia may occur if the patient stands without an anti-gravity suit upon which he has become dependent.

Phenylephrine and indirectly acting vasoconstrictor drugs like ephedrine have been used to reduce pooling, but such drugs may cause recumbent hypertension of dangerous degree with systolic pressures of 190–200 and diastolic pressures of 130. Recently there have been enthusiastic reports of the use of the combination of tyramine with a mono-aminoxidase inhibitor (Johnson *et al* 1977). We have studied these drugs on a short term basis (Fig. 13) and have failed to confirm the beneficial effects (Davies *et al* 1978) on the postural hypotension of autonomic failure. Ephedrine was more effective in this short term trial, though even ephedrine also has its risks in causing recumbent hypertension.

The most effective long term treatment is to devise means of increasing the blood volume; 9 alpha fluro-hydrocortisone not only

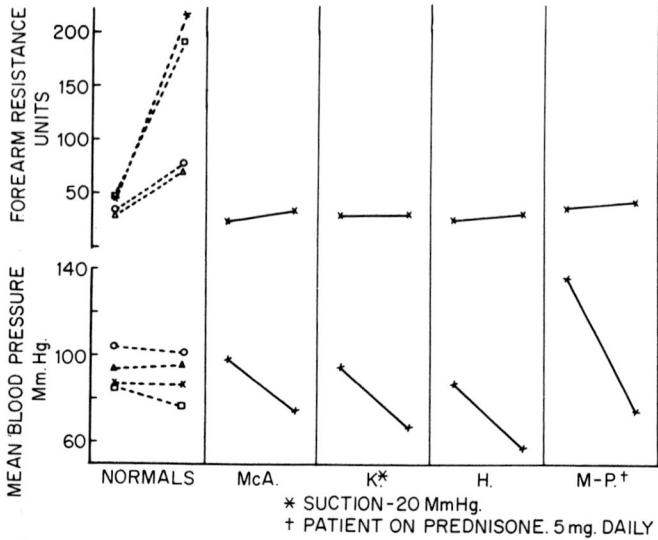

Fig. 11 The effect of suction at − 40 mm Hg (except K) on mean arterial blood pressure and forearm vascular resistance of four normal subjects and of four patients with idiopathic orthostatic hypotension. The results are presented as pairs of observations joined by a line. The first point represents the control value for forearm vascular resistance (upper half) or mean arterial blood pressure (lower half) and the second point the mean value obtained during the second minute of an exposure to suction. The results are representative of two or three periods of suction. (From Bannister *et al* 1969, used with permission.)

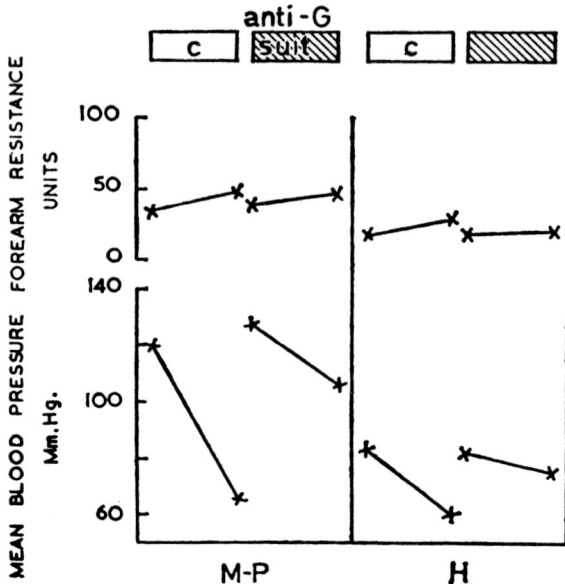

Fig. 12 The effect of an anti-gravity suit on the response to suction in idiopathic orthostatic hypotension. (From Bannister *et al* 1969, used with permission).

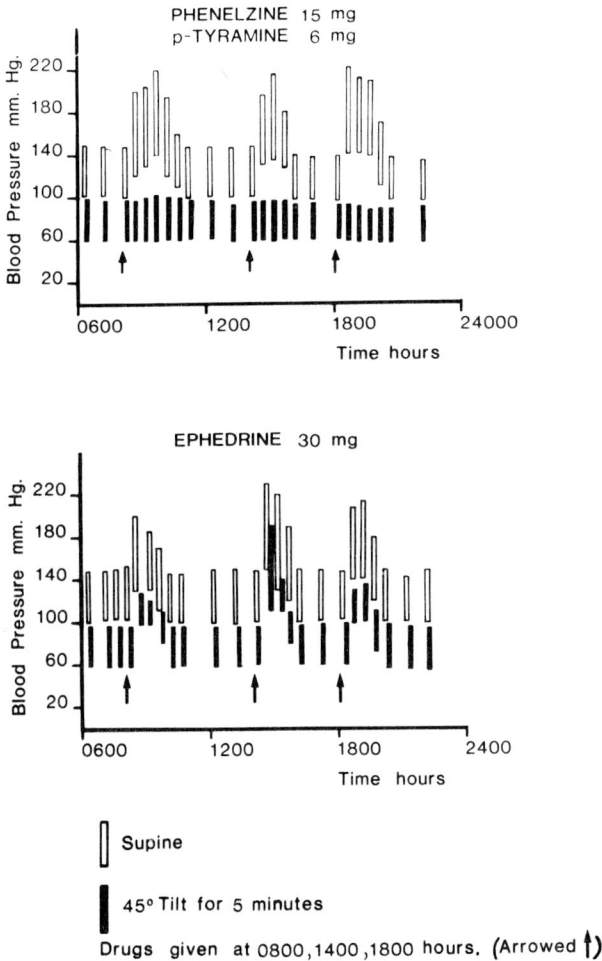

Fig. 13 The effect of tilt in patients with chronic autonomic failure after treatment with oral pressor drugs; a) phenelzine and p. tyramine; b) ephedrine. (From Davies *et al* 1978, used with permission.)

does this but in addition probably has an effect on smooth muscle in vessel walls, by sensitising them to small amounts of transmitter (Schmid *et al* 1966). Figure 14 shows the result of a trial of this drug. In one patient a mean systolic pressure at the end of body suction, before treatment with 9 alpha fluro-hydrocortisone, was converted to a normal blood pressure under these conditions after treatment. Further work is needed to establish the nature of the

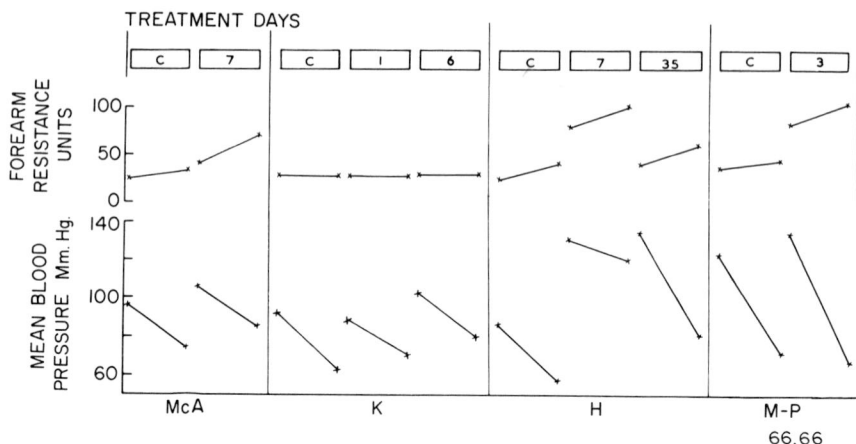

Fig. 14 The effect of 9 alpha fluorohydrocortisone on the response to suction in idiopathic orthostatic hypotension. The duration of treatment in days is indicated in the rectangle. The dose of the drug was 0.5 mg twice each day. (From Bannister *et al* 1969, with permission.)

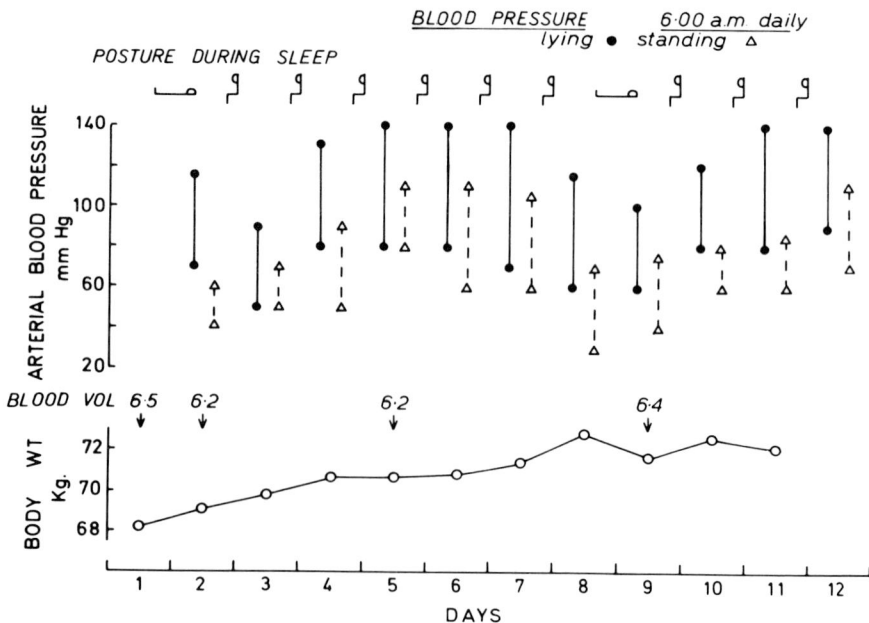

Fig. 15 The change in the early morning blood pressure (lying and standing) in a patient (H) with idiopathic orthostatic hypotension when he slept in the sitting position for 10 days with one interruption. The changes in blood volume and body weight are also shown. (From Bannister *et al* 1969, used with permission.)

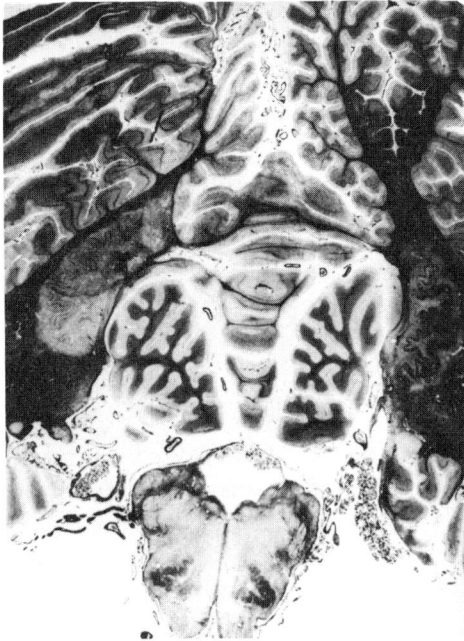

Fig. 16 Autonomic failure with multiple system atrophy. Holzer stain, showing gliosis of restiform bodies, vestibular nuclei, dorsal vagal nuclei, dorsal longitudinal fasciculi, olives, and cerebellar white matter. (From Bannister and Oppenheimer 1972, used with permission.)

differing responses of patients to 9 alpha fluro-hydrocortisone. Figure 15 shows a further method of treatment by keeping a patient in the semi-erect position at night, thereby attempting to correct the defective sodium homeostasis known to occur in autonomic failure (Wilcox *et al* 1977) This postural treatment causes an increase in body weight which must be interpreted as an increase in extracellular fluid volume due to sodium retention. After one night of lying flat both the blood pressure and the body weight fell. This method of treatment, combined with 9 alpha fluro-hydrocortisone, is probably the best way to help these patients, at least until the unfortunate stage when their other neurological disabilities overtake them.

Pathology

Dr. Oppenheimer greatly advanced knowledge of the pathology of autonomic failure by undertaking cell counts on the intermediolateral columns of the spinal cords of patients with chronic autonomic failure (Johnson *et al* 1966). Table V shows that the mean cell count is less than 20% of that of the normal controls (Bannister and Oppenheimer 1972). This 80% loss approximately reflects the reduction in resting plasma noradrenaline, and the corresponding reduction of sympathetic impulse traffic and noradrenaline release may be therefore explained.

In nearly all cases of pure autonomic failure (idiopathic orthostatic hypotension) that have been examined pathologically there is also degeneration of the substantia nigra and brain stem nuclei, particularly the dorsal nucleus of the vagus. Lewy inclusion bodies may also be found (Table VI). Other clinical and pathological features make it possible to subdivide the remaining cases into 2 further groups. In the first group with clinical features of parkinsonism there are also pathological features of idiopathic parkinsonism, which usually include Lewy inclusion bodies (Graham and Oppenheimer 1969). The second group with clinical features indicating striato-nigral or olivo-ponto-cerebellar degeneration (Figs. 16 and 17) does not have Lewy bodies. Two cases fitting this group were described originally by Shy and Drager in 1960. However because of the variable involvement of different systems the name 'multiple system atrophy' (Bannister and Oppenheimer 1972) has been suggested as more appropriate. It groups together the other structures at risk, in addition to the degeneration of brain stem nuclei.

TABLE V

Intermediolateral Cells: Mean cell counts, with standard error of mean, showing number of cells in lateral horns in 20u sections.

	Pooled Controls	Johnson *et al* 1966, case 1	Johnson *et al* 1966, case 2	Graham & Oppenheimer, 1969	Bannister & Oppenheimer, case 1	Bannister & Oppenheimer, case 2	Bannister & Oppenheimer, case 3
Upper thoracic	15.8 ± 1.0	1.3 ± 0.3	4.3 ± 0.5	4.2 ± 0.6	4.9 ± 0.4	4.3 ± 0.4	6.5 ± 0.5
Mid thoracic	9.2 ± 0.7	1.0 ± 0.2	1.9 ± 0.2	1.3 ± 0.2	1.9 ± 0.2	1.2 ± 0.2	2.5 ± 0.2
Lower thoracic	9.5 ± 0.7	2.0 ± 0.3	2.7 ± 0.3	1.3 ± 0.2	2.7 ± 0.3	2.0 ± 0.3	3.4 ± 0.4

TABLE VI

Pathological Changes in Syndromes of Autonomic Failure

Lesions	Intermedio-lateral column cells	Pigmented nuclei of brain stem	Pontine nuclei Cerebellar cortex Inferior olives	Lewy bodies
Autonomic Failure	+	+	0	+
Autonomic Failure with Parkinsonism	+	+	0	+
Autonomic Failure with Multiple System Atrophy	+	+	+	0

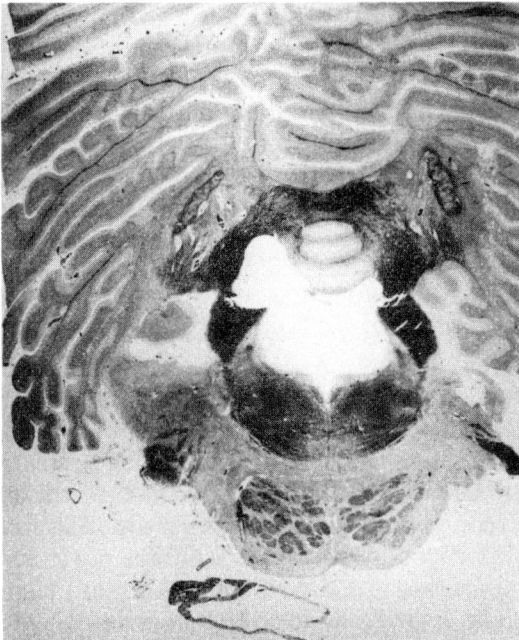

Fig. 17a. Autonomic failure with multiple system atrophy. Section of pons and cerebellum, stained for myelin. There is myelin loss in the transverse pontine bundles, middle cerebellar peduncles and cerebellar white matter, and to a lesser extent in the central tegmental tracts. (From Bannister and Oppenheimer 1972, used with permission.)

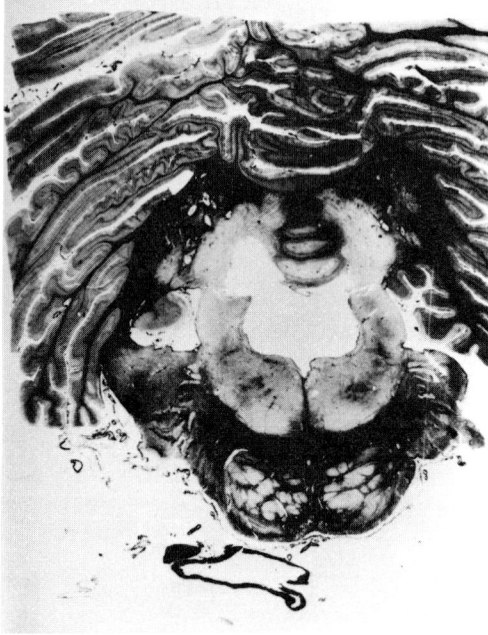

Fig. 17b Autonomic failure with multiple system atrophy. Section of pons and cerebellum, stained for fibroglia. There is gliosis in the transverse pontine bundles, middle cerebellar peduncles and cerebellar white matter, and to a lesser extent in the central tegmental tracts. (From Bannister and Oppenheimer 1972, used with permission.)

There is as yet no satisfactory hypothesis explaining this remarkable selective degeneration. In progressive autonomic failure this always affects the pre-ganglionic sympathetic neurones (the intermediolateral column cells) and usually also the pigmented brain stem nuclei. These cells are derived from the basal plate of the primitive neural tube, and contain melanin, which is an oxidation product of dopamine and noradrenaline (Bannister 1971). Melanin is thought to be the precursor of the Lewy bodies. We do not know whether the post-ganglionic defects which have been demonstrated to a variable degree distal to the intermediolateral column cells are due to transneuronal degeneration, to loss of impulse traffic, or to some biochemical factor which these cells have in common. The immediate obstacle to looking for a biochemical factor in common is that some of the cells affected are cholinergic and some are adrenergic: however the distinction between cholinergic and adrenergic cells is now no longer as precise as it was, for cholinergic transmission can be modulated by adrenergic nerve endings on these cells. Newer techniques which include histo-chemical fluorescence localisation, neurochemical transmitter assays, the administration of selective transmitters and blockers, and iontophoretic stimulation of cells within the central nervous system may well provide a better hypothesis which can be tested. In the mean-time the syndrome of chronic autonomic failure continues to provide both a challenge for treatment and a stimulus to greater understanding of the autonomic nervous system in particular and neuronal degenerations in general.

References

Aminoff, M.J. and Wilcox, C.S. (1971). *Brit. Med. J.* **4**, 80.
Bannister, R. (1971). *Lancet* **2**, 175.
Bannister, R., Ardill, L. and Fentem, P. (1967). *Brain* **90**, 725.
Bannister, R., Ardill, L. and Fentem, P. (1969). *Q. J. of Med.* **38**, 377.
Bannister, R. and Oppenheimer, D.R. (1972). *Brain* **95**, 457.
Bannister, R., Sever, P. and Gross, M. (1977). *Brain* **100**, 327.
Barany, F.R. and Cooper, E.H. (1956). *Clin. Sci.* **15**, 533.
Bradbury, S. and Eggleston, C. (1925). *Am. Ht. J.* **1**, 73.
Davies, B., Bannister, R. and Sever, P. (1978). *Lancet* **1**, 175.
Fichefet, J.P., Sternon, J.E., Franken, L., Demanet, J.C. and Vanderhaeghen, J.J. (1965). *Acta Cardiol.* **20**, 332.
Folkow, B. and Neil, E. (1971). Circulation. Oxford University Press.
Fullerton, P.M. and Barnes, J.M. (1966). *Br. J. Ind. Med.* **25**, 210.
Gray, I.R. (1972). *Brit. Med. J.* **1**, 32.
Graham, J.G. and Oppenheimer, D.R. (1969). *J. Neurol., Neurosurg., Psychiat.* **32**, 28.
Hopkins, A., Neville, B. and Bannister, R. (1974). *Lancet* **1**, 769.
Johnson, R.H., Lee, G. de J., Oppenheimer, D.R. and Spalding, J.M.K. (1966). *Q. J. Med. N.S.* **35**, 296.
Johnson, R.H. and Spalding, J.M.K. (1974). Disorders of the Autonomic Nervous System. Blackwell.
Low, P.A., Walsh, J.C., Huang, C.Y. and McLeod, J.G. (1975). *Brain* **98**, 341.
Mathias, C.J., Matthews, W.B., Spalding, J.M.K. (1977). *J. Neurol., Neurosurg., Psychiat.* **40**, 138.
McLeod, J.G. and Penny, R. (1969). *J. Neurol., Neurosurg., Psychiat.* **32**, 297.
Nanda, R.M., Johnson, R.H. and Keogh, H.J. (1976). *Lancet* **2**, 1164.
Shy, G.M. and Drager, G.A. (1960). *Arch. Neurol.* (Chicago) **2**, 571.
Schmid, P.G., Eckstein, J.W., Abboud, F.M. (1966). *Circulation* **34**, 620.
Tomashefsky, A.J., Horwitz, S.J. and Feingold, M.H. (1972). *Neurology* (Minneap.) **22**, 251.
Wheeler, T. and Watkins, P.J. (1972). *Brit. Med. J.* **2**, 584.
Wilcox, C.S., Aminoff, M.J., Kurtz, A.B. and Slater, J.D.H. (1974). *Clin. Sci. Mol. Med.* **46**, 481.
Wilcox, C.S., Aminoff, M.J. and Slater, J.D.H. (1977). *Clin. Sci. Mol. Med.* **53**, 321.

THE PUPIL AS AN INDICATOR OF DENERVATION HYPERSENSITIVITY IN THE AUTONOMIC NERVOUS SYSTEM

C. PALLIS

Royal Postgraduate Medical School, Du Cane Road, London, U.K.

Introduction

The size of the human pupil and the pattern of its response to locally applied pharmacological agents may provide insights of clinical value. Such observations also help document, in a readily visible manner, some interesting disorders of neurotransmitter function.

Neuropharmacological manipulations may help distinguish between preganglionic and postganglionic lesions of the autonomic innervation of the musculature of the iris. The relevant tests are based on the detection of denervation hypersensitivity and are used to differentiate:

a) patients with ptosis and miosis due to a 'third neurone' lesion (vide infra) of the *sympathetic* innervation of the dilator pupillae from patients in whom Horner's syndrome arises from more proximal involvement of the pupillodilator pathway.

b) patients with Adie's syndrome (currently thought to be due to a lesion of postganglionic *parasympathetic* fibres to the sphincter pupillae) from patients with large pupils due to preganglionic lesions of the pupilloconstrictor fibres.

Although neuropharmacological investigation of the pupil may be diagnostically rewarding, the findings should always be interpreted in the light of established anatomical and physiological facts. A recapitulation of these facts may be helpful in establishing the circumstances under which sympathetic or parasympathetic denervation hypersensitivity might be anticipated. Some comments are also warranted concerning the physiology of denervation hypersensitivity in the autonomic nervous system.

Anatomy

Sympathetic Innervation of the Dilator Pupillae

The 'outflow pathway' extends from the hypothalamus down through the brain-stem and cervical cord, then out of the spinal canal with ventral roots of (C8)-T1-(T2), and up again into the skull.

For purpose of convenience the segment between the hypothalamus and the cervical ciliospinal centre (vide infra) is described as the 'first neurone' (although it is probably interrupted by several synapses in the pontine and mesencephalic tegmentum). The 'second neurone' extends from the ciliospinal center to the superior cervical ganglion. The 'third neurone' travels from the superior cervical ganglion to the dilator pupillae.

Preganglionic fibres ('second neurone') The cell bodies lie in the intermediolateral grey columns of the lowest cervical and uppermost thoracic segments of the spinal cord, where they constitute the so-called 'cilio-spinal centre of Budge'.

In man most preganglionic fibres destined for the eye leave the spinal cord with the ventral roots of the first thoracic segment. Small contingents may also travel via the C8 and T2 roots. From here the fibres gain access (via the white rami communicantes) to the paravertebral sympathetic chain. They then proceed upwards, without synapsing, through the inferior and middle cervical ganglia, eventually reaching the superior cervical ganglion.

The superior cervical ganglion, which respresents a fusion of the first four cervical sympathetic ganglia, lies between the internal jugular vein and the internal carotid artery, below the base of the skull (i.e. rather higher than is usually thought). Both oculo-sympathetic and sudomotor fibres destined for the face synapse in it.

Postganglionic fibres ('third neurone') The pupillodilator fibres leave the ganglion and accompany the internal carotid artery through the carotid canal and foramen lacerum and into the region of the Gasserian ganglion. (Some fibres contribute, en route, to the tympanic plexus on the promontory of the middle ear, before re-joining the internal carotid plexus). The oculo-sympathetic fibres lie in close apposition to the internal carotid artery in the cavernous sinus. Most of them join the ophthalmic division of the trigeminal nerve, entering the orbit with its nasociliary branch. The *long ciliary nerves* leave this branch, by-pass the ciliary ganglion, penetrate the sclera and choroid (both nasally and temporally) and eventually reach the dilator pupillae.

Postganglionic sympathetic fibres are also distributed to other ocular structures. Those innervating blood vessels or uveal chroma-tophores (including those of the iris) share the early part of the postganglionic pathway. They leave the nasociliary nerve as the *'long roots' of the ciliary ganglia* and pass through these structures (without synapsing) en route to their effector organs.

Most sudomotor and piloerector postganglionic fibres to the face leave the superior cervical ganglion and reach their destination via plexuses along the external carotid artery and its branches. Sudomotor fibres to the forehead may re-enter the skull and travel much of their course with the pupillodilator fibres, eventually reaching their target via the ophthalmic artery and its supraorbital branch.

It will be seen from this recapitulation that if a patient with ptosis and miosis has neuropharmacological evidence of an associated denervation hypersensitivity of his constricted pupil, the lesion is unlikely to be in the brain-stem, cervical segments of the spinal cord, ventral roots or ascending sympathetic chain in the neck. It is likely to be at the base of the skull, along the course of the internal carotid artery, in the middle ear cleft, in the cavernous sinus, or in the orbit.

Parasympathetic Innervation of the Sphincter Pupillae

The 'outflow pathway' to the sphincter pupillae is via a two neurone system.

First (preganglionic) neurone This arises in the Edinger-Westphal nucleus in the rostral midbrain. It travels via the third cranial nerve, its branch to the inferior oblique muscle and the short root of the ciliary ganglion. This ganglion is situated in the loose fatty tissue at the apex of the orbit, between the optic nerve and the lateral rectus muscle.

Second (postganglionic) neurone This arises from cell bodies in the ciliary ganglion. Fibres travel via the *short ciliary nerves* and are distributed to the sphincter pupillae. On their way there these fibres pierce the sclera at the posterior pole of the globe of the eye. They then run anteriorly, first in the sclera itself, later in the plexus in the subchoroid space. Lesions at these sites are more common than most neurologists think. Most such patients are seen by ophthalmologists.

All pupilloconstrictor fibres probably reach the iris via a synapse in the ciliary ganglion. According to Walsh and Hoyt (1969) there is no anatomical basis for the idea of cholinergic pupilloconstrictor fibres that bypass the ciliary ganglion, or that synapse in the episcleral cells which are sometimes found along the course of the short ciliary nerves. The existence of such an alternative pathway had been hypothesised (Nathan and Turner, 1942) to account for isolated light rigidity of the pupil following peripheral injury.

It is important to appreciate that the vast majority (94%) of parasympathetic postganglionic fibres leaving the ciliary ganglion are *not* concerned with pupillary constriction (Warwick, 1954). They are distributed to the ciliary muscle and are concerned with accommodation. This observation is crucial to the modern under-standing of the pathogenesis of Adie's syndrome.

It will be seen from these anatomical facts that if a patient with mydriasis has an associated denervation hypersensitivity of his dilated pupil the lesion is unlikely to be in the brain-stem, in the cavernous sinus or in the superior orbital fissure. The lesion is likely to be at the orbital apex (i.e. in the ciliary ganglion) or in the outer layers of the eye itself.

Physiology

Degeneration of a postganglionic fibre is followed by enhanced responsiveness of the effector cell to the neurotransmitter normally liberated at the end of the fibre concerned. This has been known for a long time as Cannon's law of denervation. There is often also enhanced responsiveness to chemically related substances.

Following postganglionic *parasympathetic* lesions such denervation hypersensitivity may appear within 48 hours. With postganglionic *sympathetic* lesions it takes longer (usually up to 3 weeks) to develop fully (Walsh and Hoyt, 1969). Its manifestations tend slowly to diminish with the passage of time, probably as a result of axonal regeneration. Parasympathetic denervation hypersensitivity of the iris may however last for many years (25 years in a patient reported by Laties and Scheie, 1965).

Sprouting from undamaged postganglionic fibres into the sheaths of adjacent degenerated fibres also contributes substantially to functional recovery in postganglionic lesions of the autonomic nervous system. When the functions subserved by such intact adjacent postganglionic fibres differ from those of the damaged fibres the *aberrant regeneration* may be associated with a wide variety of clinically intriguing phenomena, including several related to the pupil. For instance aberrant regeneration between the numerous post-ganglionic fibres to the ciliary muscle and the scanty constrictor fibres to the pupil accounts for some of the features of Adie's syndrome.

Pupillary Responses to Topically Instilled Drugs

Before inferences are drawn from clinical tests some methodological points should be stressed concerning testing procedure and some comments made about normal responses.

a) The technique of instillation is important. Both lacrimation and pressure or squeezing of the eyelids should be avoided. Each 'instillation' should consist in the administration of a single drop, followed after 2 minutes by a further drop. Between the two applications the patient should be asked gently to close the eye under investigation and to 'roll it around'. Three such paired instillations at intervals of ten minutes (i.e. a total of 6 drops to each eye) may have to be administered before a firm opinion can be given as to whether there has been a response or not.

b) It has been known for a long time that in normal subjects the responses may occasionally be delayed (Schultz, 1908). No definite conclusion to the effect that no reaction has taken place should therefore be drawn before 1 hour has elapsed. Although both pupils may eventually show a quantitatively similar response, one eye may react somewhat earlier than the other. This phenomenon is more likely to be due to differences in the rates of penetration of drugs or in the amounts delivered than to subclinical lesions of autonomic pathways.

c) There are variations between normal individuals in the sensitivity of the iris musculature to topically applied drugs. It is not yet known for certain whether these reflect differences in drug penetration or true differences in muscle sensitivity.

On the whole Caucasians are more responsive to mydriatics than Chinese, and the latter more responsive than Negroes (Chen and Poth, 1929). Among Caucasians, mydriatics are more effective in individuals with light irides than in those with dark (Howard and Lee, 1927; Obianwu and Rand, 1965). Such variations have usually been reported in relation to 'adrenergic' mydriatics, less often in relation to drugs dilating the pupils by interfering with cholinergic transmission. At the concentrations used in clinical testing (0.1% adrenaline) such racial differences will not of course produce false positive results, although they may possibly account for some false negatives.

Less is known about differential responses to miotics in relation to the degree of iris pigmentation, although Melikian *et al* (1971) showed that persons with heavily pigmented skin, irides, trabecular mesh work and fundi were less responsive to the pressure and outflow facility effects of 4% pilocarpine than those with lightly pigmented structures.

d) In the presence of a suspected pupillary abnormality the neuro-pharmacological investigation should be carefully planned to avoid false positive and false negative results.

Unilaterally abnormal pupil It is essential to test both pupils if false positive reactions are to be avoided. For instance, in a patient with unilateral mydriasis slight contraction may occur after the instillation of 2.5% mecholyl without this implying denervation hypersensitivity. Only when the dilated pupil contracts while the other pupil fails to respond has true denervation hypersensitivity been demonstrated.

It is also important to detect false negative reactions. If the clinical situation is strongly suggestive of Adie's syndrome and the appropriate test (with 2.5% mecholyl) fails to show evidence of denervation hypersensitivity it is permissible to use stronger solutions (say 5% and 10%) provided both eyes are tested. Only when the normal pupil begins to contract without a more extensive constriction of the dilated pupil can denervation hypersensitivity of the dilated pupil be excluded.

Bilaterally abnormal pupils No such comparisons are possible. Only one eye should be tested at a time, the other being used as a control.

e) The general state of the patient influences the size of the pupils. Apprehensive patients have wider pupils than drowsy ones.

Testing for Sympathetic Denervation Hypersensitivity: Responses to Adrenaline and Cocaine

The 0.1% Adrenaline Test

Noradrenaline is the neurotransmitter involved at sympathetic terminals in the iris. It is however adrenaline that is used in these clinical tests, a) because it is more stable, and b) because it is a more potent mydriatic (von Euler, 1945; Burn and Hutcheon, 1949; Howell, 1934 compared with Owe-Larsson, 1956).

In adequate concentration adrenaline is capable of directly stimulating the receptor sites of the dilator pupillae. However, the normal pupil will not dilate in response to the instillation of 0.1% (1 in 1000) adrenaline drops.* In the presence of denervation hyper-

---

*Stronger adrenaline solutions may also fail to dilate the normal pupil. Willets (1969) using freshly prepared 1% solutions of 1-adrenaline bitartrate and 1-noradrenaline bitartrate (and also 2% solutions of 1-adrenaline hydrochloride) found that none of these concentrations had a mydriatic effect when tested on 18 healthy volunteers at the Johns Hopkins Hospital. The technique of administration differed however from the one recommended. The drops

sensitivity this concentration of adrenaline will produce obvious mydriasis. Within limits, the extent of dilatation will reflect the degree of denervation. Walsh and Hoyt (1969) stress that 'theoretically supersensitivity should be greatest when the postganglionic neurone is destroyed, less when the preganglionic neurone is interrupted and even less when the central sympathetic pathway is involved'.

Patients with lesions of the third neurone of the oculosympathetic pathway may exhibit unequivocal mydriasis (i.e. dilatation of more than 2 mm) in response to such a stimulus. Jaffe (1950) describes this as 'a very reliable test' but Walsh and Hoyt (1969) found the response less convincing. If the lesion affects the second neurone there may also be slight dilatation. According to Jaffe (1950) 'sensitivity with such a lesion is about one-half that seen after postganglionic sympathectomy'. There is widespread agreement that lesions of the first neurone are not associated with this pattern of response, although even here there are one or two exceptions (Jaffe, 1950 – case 2; Walsh, 1957).

Personal observations

Normals Our own experience, based on a series of 50 normal volunteers (25 men and 25 women in the 20–40 age group), tested in the way described, confirms the view that 0.1% adrenaline solutions, used alone, have no mydriatic effect. False positive diagnoses of sympathetic denervation hypersensitivity should therefore not occur.

First neurone lesions We have tested 12 patients with Horner's syndrome attributable to such a lesion. There were 5 patients with lateral medullary infarction, 2 with pontine glioma, 1 with a pontine tuberculoma, 3 with cervical syringomyelia and 1 with an intra-medullary cervical neoplasm. The 0.1% adrenaline test was negative in all instances.

Second neurone lesions The results here were also overwhelmingly negative. 18 patients were tested. 12 had malignant disease at the root of the neck (either primary or secondary), 5 were suffering from the sequelae of trauma (including surgical or obstetric trauma) and 1 had an asymptomatic paravertebral neurofibroma. One of the patients with signs attributable to trauma had had bilateral pre-

---

were instilled into one eye (using the other as a control) at 8 hourly intervals for a period of 4 to 5 days. Examinations were repeatedly carried out at intervals varying from 30 minutes to 7 hours after the most recent instillation.

ganglionic sympathectomies 30 years earlier for Raynaud's disease, resulting in a striking bilateral Horner's syndrome.

The 0.1% adrenaline test was negative in 16 of these 18 cases. A 2mm mydriasis occurred in the patient with neurofibromatosis (in whom a more distal lesion could not be excluded with certainty). The patient with bilateral sympathectomies showed similar slight dilatation on one side (the side with most ptosis) but none on the other. The significance of this is uncertain.

Third neurone lesions Seven patients were tested. One had a metastasis from nasopharyngeal carcinoma involving the base of the skull: the Horner's syndrome was associated with ipsilateral involvement of the lower three cranial nerves. One patient had an orbital granuloma (Tolosa Hunt syndrome) extending through the superior orbital fissure into the cavernous sinus. Two patients had aneurysms of the internal carotid artery in the cavernous sinus and exhibited the unusual association of external ophthalmoplegia and small pupils. Three were thought to be suffering from that variant of Raeder's syndrome in which no structural lesion can be demonstrated. They had paroxysmal unilateral headaches with ptosis and miosis — but no anidrosis — and a normal carotid angiogram.

In four of the patients the 0.1% adrenaline test was unequivocally positive (Fig. 1). In three (the first two patients mentioned and one of the patients with Raeder's syndrome) the test gave equivocal results (dilatation of less than 2mm).

The Cocaine Test

Cocaine dilates the normal pupil by blocking catecholamine reuptake at adrenergic endings. The instillation of 4% cocaine eyedrops will dilate the normal pupil by this mechanism. Cocaine will only exhibit this effect, however, if adequate amounts of noradrenaline are being released at the effector junction between the postganglionic fibre and the dilator pupillae.

Used alone, 4% cocaine will fail to dilate the miotic pupil of a patient with a third neurone lesion. With second neurone lesions there is also often an impaired or absent mydriatic effect (Jaffe, 1950; Walsh and Hoyt, 1969) — an observation which raised interesting speculations on the effect of a second neurone lesion on the release of neurotransmitter more distally. Patients with first neurone lesions usually show full pupillary dilatation to the cocaine test, for reasons which are not fully clear.

Because of these overlapping results — and because of slight

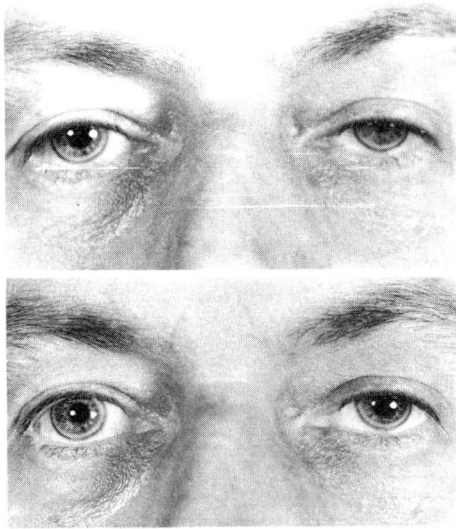

Fig. 1. 55 year old man with 6 months history of recurrent left hemicranial and retro-ocular headaches. Examination (3 days after a headache) revealed left ptosis and miosis but no other neurological signs. Skull x-rays: normal. Left carotid angiogram: normal. No anidrosis over left side of face. 0.1% adrenaline drops, instilled into both eyes, produced dilatation on the left. *Diagnosis*: Raeder's syndrome.

variations in response in patients with second or third neurone lesions according to the completeness or otherwise of the lesion — the cocaine test is seldom used alone. In isolation it is seldom of value in establishing the precise site of an oculosympathetic lesion. At best it

TABLE I

Pupillary dilatation in response to instilled drugs

	0.1% adrenaline	4% cocaine
Normal	0	++
1st neurone lesion	0	+
2nd neurone lesion		
a) minor	+/−	+/−
b) major	+	0
3rd neurone lesion		
a) minor	+	+/−
b) major	++	0

can help to establish or corroborate, through the failure of normal mydriasis, the presence of such a lesion in a clinically suspect case. One advantage of the test is that it may be of diagnostic value in very recent lesions, which is not the case with the 0.1% adrenaline test, which depends on denervation hypersensitivity. This, as has previously been mentioned, may take a few days to develop.

Stronger (say, 10%) solutions of cocaine should not be used. They may anaesthetise the nerves to the sphincter, thereby causing mydriasis and vitiating the interpretation of results.

The table summarises our experience with these two tests and is in general agreement with the findings of Jaffe (1950) and of Walsh and Hoyt (1969).

The Combined Test

Denervation hypersensitivity of the receptors of the dilator fibres is best demonstrated by using the 0.1% adrenaline test afer pre-treatment with 2—4% cocaine eye drops. The cocaine enhances corneal permeability, ensuring that more adrenaline reaches the receptor. It then enhances adrenergic effects at the appropriate sites.

In practice the test is carried out by instilling two drops of 4% cocaine into each eye, on three occasions if necessary, as previously described in the section concerning technique. A positive response (i.e. unequivocal mydriasis) suggests a first neurone lesion. If there is no response (or if the response is equivocal) 0.1% adrenaline drops are instilled half an hour later, according to a similar schedule. If there is then only slight dilatation a second neurone lesion is probable. Unequivocal dilatation is diagnostic of a third neurone lesion. Such a sequence was observed in all three of the aforemention-ed patients with suspected third neurone lesions, in whom tests with 0.1% adrenaline alone were equivocal.

It is sometimes said that although an unequivocally positive adrenaline test is of considerable diagnostic value a negative response (even after pre-treatment with cocaine) does not exclude post-ganglionic involvement of the ocular sympathetic. I have not encountered such a case.

Heterochromia Iridis

During embryological development the melanophores travel into the iris and choroid under sympathetic influence. The sympathetic nervous system is one of the factors influencing the development of melanin pigment and therefore iris colour. In the absence of sympathetic input the iris may remain hypopigmented.

Failure of one iris to pigment in a normal manner may be associated with injury to the sympathetic fibres early in life (for instance birth injury to the brachial plexus). Depigmentation after oculosympathetic injury in adults is very rare, although occasional well documented cases have been reported (Makley and Abbot, 1965; Walsh and Hoyt, 1969; Ehinger *et al*, 1969). The occurrence of this depigmentation suggests the persistence into adult life of some kind of sympathetic trophic influence on the melanocytes.

Heterochromia iridis may result from obvious local disease, when problems may arise in deciding which is the abnormal side (one need only think of the different effects of diffuse iris melanoma on the one hand and of iris atrophy following iritis or glaucoma on the other). It may also occur as an isolated congenital anomaly or be associated with other congenital anomalies, not necessarily involving sympathetic pathways. Finally it may be noted in association with other evidence of oculosympathetic palsy, such cases usually being described in the literature as congenital Horner's syndrome. In two personally observed patients with congenital Horner's syndrome (no evidence of birth injury) with anisochromia iridis the miotic pupil showed unequivocal evidence of denervation hypersensitivity.

Testing for Parasympathetic Denervation Hypersensitivity: Responses to Mecholyl

The normal neurotransmitter acting at postganglionic parasympathetic endings in the sphincter pupillae is acetylcholine. This substance is however unsuitable for clinical use in the form of eye drops. Mecholyl (acetyl-beta-methylcholine) is structurally similar to acetylcholine and is capable of depolarising the effector cell, causing the pupil to contract.

A patient with an isolated large pupil which reacts poorly to light and in a slow, tonic manner on looking at a near object, is said to have Adie's syndrome (for an excellent review of the whole subject see Loewenfeld and Thompson, 1967). Adler and Sheie (1940) first suggested that the physiological disturbance in this condition was 'a partial denervation of the parasympathetic supply to the pupil at, or peripheral to, the ciliary ganglion'. They also introduced the use of 2.5% mecholyl eye drops into clinical diagnosis.

These authors investigated 11 patients with Adie's syndrome, instilling one drop of 2.5% mecholyl into each eye and repeating the procedure after 5 minutes. In each instance the tonic pupil showed an intense miosis whereas unaffected pupils showed no response. In 48 normal controls the authors found that mecholyl drops at various concentrations below 15% had no effect 'other than a slight occasional anisocoria which could well be accounted for by the irritation and hyperaemia produced by the drug'. This was distinguishable from 'true miosis'.

Loewenfeld and Oono (1966) described similar experiments but were more guarded in their interpretation. They found that in several of their 27 normal subjects (in whom instillation of 2.5% mecholyl solution was given in the form of drops, with a two minute interval between the first and second drop) the pupil reacted 'slightly'. They confirmed the occasional unresponsiveness to much higher concentrations. Some individuals failed to respond, even when tested with a 25% solution. They stressed that false positive and false negative responses to the mecholyl test could be avoided if attention was paid to the previously mentioned methodology.

Our experience confirms this assessment. In 18 out of 20 normal subjects (10 males and 10 females, aged between 20 and 40) there was no significant pupillary response to three instillations (6 drops) of 2.5% mecholyl. In two subjects there was a response in the micro-semeiological range (1 mm of miosis or less).

Our experience with Adie's syndrome is limited. Three out of four personally tested cases showed unequivocally positive responses to 2.5% mecholyl. The remaining patient showed a unilateral positive response to 10% mecholyl, on the side of the dilated pupil.

The attribution of a peripheral aetiology to a case of internal ophthalmoplegia is usually an easy matter and does not require a search for denervation hypersensitivity. Most neurologists will have seen patients with internal ophthalmoplegia following herpes zoster ophthalmicus. I have seen three such cases and all showed a positive mecholyl test. Patients with metastic tumours of the choroid (Liegl and Kohn, 1962) or patients who have been subjected to trans-scleral diathermy for retinal detachment (Kronfeld, 1961) may also develop internal ophthalmoplegia with denervation hypersensitivity but such cases will seldom be seen by neurologists.

A negative mecholyl test has proved diagnostically useful on at least one occasion. The patient was a child with isolated bilateral internal ophthalmoplegia of central origin who eventually turned out to have a pineal tumour. The mecholyl test was repeatedly negative (even using a 10% solution). This encouraged a persistent, and eventually successful, search for a central cause.

It has been claimed that mecholyl solutions are unstable and that, when stored, variations in the response to mecholyl drops, believed to be of a given concentration, could be due to variations in the effective concentration of the miotic agent. Loewenfeld and Oono (1966) investigated this problem, using 10% mecholyl in an aqueous solution. They concluded that variations in response were not due to alterations in the miotic agent. Provided the solution was stored in a refrigerator at 41°F there was no noticeable reduction in its miotic potency in the same subject over a period of one year. Differences between individuals, and differences between different solutions supposedly of the same strength, far outweighed differences in responsiveness of the same individual to the same solution within the period of the experiment.

The practical implication of all this is that in testing for cholinergic sensitivity in the diagnosis of Adie's syndrome it is neither necessary nor advisable to make up a fresh solution of mecholyl for each patient.

References

Adler, F.H. and Scheie, H. (1940). *Trans. Amer. Ophthal. Soc.* **38**, 183.

Burn, J.H. and Hutcheon, D.E. (1949). *Brit. J. Pharmacol.* **4**, 373.

Chen, K.K. and Poth, E.J. (1929). *J. Pharm. Exp. Therap.* **36**, 429.

Ehinger, B., Falck, B. and Rosengren, E. (1969). *Arch. Clin. Exp. Ophthalmol.* **177**, 206.

Howard, H.J. and Lee, T.P. (1927). *Proc. Soc. Exp. Biol. Med.* **24**, 700.

Howell, S.C. (1934). *Arch. Ophthal.* **12**, 833.

Jaffe, N.S. (1950). *Arch. Ophthal.* **44**, 710.

Kronfeld, P.C. (1961). *Trans. Amer. Ophthal. Soc.* **59**, 239.

Laties, A.M. and Scheie, H.G. (1965). *Arch. Ophthal.* **74**, 458.

Liege, O. and Kohn, K. (1962). *Clin. Mbl. Augenheilk.* **140**, 327.

Loewenfeld, I.E. and Oono, S. (1966). *Ear Nose Throat Monthly* **45**, 69.

Loewenfeld, I.E. and Thompson, H.S. (1967). *Amer. J. Ophthal.* **63**, 46.

Makley, T.A. and Abbott, K. (1965). *Amer. J. Ophthal.* **59**, 927.

Melikian, H.E., Lieberman, T.W. and Leopold, I.H. (1971). *Amer. J. Ophthal.* **72**, 70.

Nathan, P.W. and Turner, J.W.A. (1942). *Brain* **65**, 333.

Ohianwu, H.O. and Rand, M.J. (1965). *Brit. J. Ophthal.* **49**, 264.

Owe-Larsson, A. (1956). *Acta Ophthal.* **34**, 27.

Schultz, W.H. (1908). *Proc. Soc. Exp. Biol. Med.* **6**, 23.

von Euler, U.S. (1946). *Acta Physiol. Scand.* **12**, 73.

Walsh, F.B. (1957). *Clinical Neuro-ophthalmology.* Second edition. Williams and Wilkins, Baltimore.

Walsh, F.B. and Hoyt, W.F. (1969). *Clinical Neuro-ophthalmology.* Third edition. Williams and Wilkins, Baltimore.

Warwick, R. (1954). *J. Anat.* **88**, 71.

Willetts, G.S. (1969). *Amer. J. Ophthal.* **68**, 216.

OTHER NEUROTRANSMITTERS IN PARKINSON'S DISEASE

CHRIS PYCOCK

Department of Pharmacology, The Medical School, University of Bristol, Bristol, U.K.

Dopamine in Parkinson's Disease

It is well established that Parkinson's disease is related to loss of dopaminergic function in the brain; the disorder is characterised chemically by gross deficits in the levels of dopamine and its metabolite homovanillic acid and of the major synthesizing enzyme tyrosine hydroxylase (Hornykiewicz, 1966, 1972; McGeer and McGeer, 1976). Two major ascending dopaminergic pathways have been described in animal brains (Ungerstedt, 1971a), the first of which is the nigro-neostriatal tract, arising from the melanin-containing cell bodies located in the zona compacta region of the substantia nigra and projecting forwards to innervate the neostriatal complex (caudate nucleus and putamen). The basic biochemical and histological pathology of Parkinson's disease is consistent with the degeneration of this dopaminergic pathway, whose major function is believed to be the initiation and control of movement.

Recently more attention has been directed towards a second ascending dopaminergic system, the so-called mesolimbic pathway. The cell bodies of this system are located caudally and medially to the substantia nigra in the region of the interpeduncular nucleus of the midbrain (designated the A10 group of cell bodies from animal studies (Ungerstedt, 1971a)). The mesolimbic pathway innervates areas of the limbic forebrain, including the nucleus accumbens septi, tuberculum olfactorium, and amygdaloid nuclei, and also projects to the frontal cortex. The functional role of this pathway is believed to be in the control of mood and behaviour, and it is the mesolimbic forebrain dopamine system which is currently thought to be associated with psychotic disturbances such as the schizophrenias. However animal studies have shown that areas of the limbic forebrain, in particular the nucleus accumbens, are also associated with motor activity (see below). A recent survey has demonstrated marked decreases in the dopamine content of the limbic parolfactory gyrus

and nucleus accumbens in cases of Parkinson's disease, in addition to
the established reduction of dopamine in the neostriatum (Farley
et al, 1977). This deficit of dopamine from limbic areas may be
associated with the akinesia and also the disturbances of mood which
are not uncommon in Parkinson's disease. Such findings now
implicate both nigro-neostrial and mesolimbic dopamine systems
in the pathology of this disorder.

Other Neurotransmitters in Parkinson's Disease

Provided with such consistent pathological and neurochemical
evidence, dopamine is accepted as being the major neurotransmitter
involved in Parkinson's disease. The beneficial effects of the dopamine
agonists in the clinic only serve to strengthen this assumption.
However, in the enthusiasm to implicate the dopamine system, the
possible role of other central neurotransmitters in Parkinson's disease
has received much less attention. Pathologically, in addition to the
characteristic loss of melanin from the substantia nigra, the parkinsonian brain often reveals degeneration of the pigmented cell bodies
from the locus coeruleus and other brain stem nuclei, as well as
lesions in the neostriatum and globus pallidus. Such additional
neuronal degeneration might implicate other neurotransmitters in
Parkinson's disease, although, with respect to striatal and pallidal cell
loss, it is not always possible to differentiate between a primary lesion
and the results of secondary degeneration due to loss of the nigro-
neostriatal dopaminergic input.

Likewise, neurochemical studies have implied the possible role of
other neurotransmitters in Parkinson's disease. 5-hydroxytryptamine
(5-HT) is found in high concentrations within normal basal ganglia,
and Hornykiewicz (1972) has recorded decreases of this monoamine
in Parkinson's disease. For example, some 50% loss of 5-HT was
observed in the striatum and substantia nigra, and a mean 20%
reduction in the globus pallidus.

The most severely affected noradrenaline (NA)-containing areas in
this disorder belong to a system composed of the paranigralis-
parabrachialis-pigmentosus area, the hypothalamus, the paramedian
thalamic cell groups and the nucleus accumbens and olfactory areas
of the limbic forebrain (Bernheimer *et al*, 1963; Farley and
Hornykiewicz, 1976). The patter of NA loss closely parallels the
'ventral periventricular NA system' recently described in rat brain
(Lindvall and Bjorklund, 1974), and probably reflects the loss of a
number of pigmented NA-containing cell bodies in the brain stem.

Acetylcholine (ACh), together with its synthesizing enzyme choline acetyltransferase (CAT) and metabolic enzyme acetylcholinesterase (AChE), are found in high concentrations within regions of the basal ganglia. McGeer and McGeer (1976) reported no significant changes in the activity of CAT in extrapyramidal nuclei in parkinsonian cases, although a mysterious doubling of activity of this enzyme was observed in the nucleus accumbens. Similarly these authors showed no significant changes in AChE activity in the areas studied, although a 25% reduction was noted in the globus pallidus and a 20% decrease had been previously found in the caudate nucleus and putamen (Rinne *et al*, 1973).

Of the putative central amino acid neurotransmitters, γ-aminobutyric acid (GABA) is probably the best-studied candidate. Current evidence suggests that GABA dysfunction may be an important aspect of certain neurological disorders such as Huntington's chorea (Bird and Iversen, 1974). Loss of basal ganglia GABA function has also been implicated in Parkinson's disease, where substantial falls of glutamate decarboxylase (GAD), the enzyme which converts glutamic acid into GABA, have been reported in all regions of the basal ganglia and in the nucleus accumbens (Hornykiewicz *et al*, 1976; McGeer and McGeer, 1976). Only slightly subnormal GAD levels were observed in other brain regions, and no apparent regional changes in GABA concentrations (although the latter may be obscured by post-mortem effects) (Hornykiewicz *et al*, 1976).

The possible involvement of these other neurotransmitter systems in Parkinson's disease is illustrated in Fig. 1. The available data implicate not only dopamine but also many of the other transmitters located in high concentrations within the basal ganglia. The significance of the changes in activity of other neurotransmitters for the symptomatology of Parkinson's disease remains at present uncertain. The possible functional roles and interactions with the dopamine systems of the brain are considered in the following sections.

Animal Models of Central Dopamine Function and Parkinson's Disease

Manipulation of central dopamine function in animals induces profound characteristic alterations in behaviour which are readily reproducible and quantifiable using numerical scoring systems. These behaviours have, rightly or wrongly, almost exclusively been classified in terms of indices of dopaminergic transmission, and the ability of an agent to produce these behavioural patterns is usually taken as a reflection of its action on dopaminergic mechanisms.

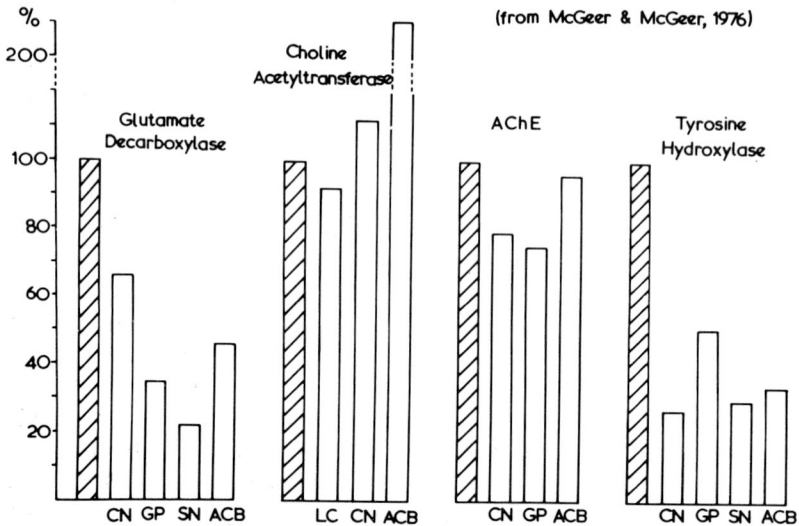

Fig. 1   Relative regional activity of the enzymes glutamate decarboxylase, choline acetyltransferase, acetylcholinesterase (AChE) and tyrosine hydroxylase in brains of patients dying of Parkinson's disease. Control activity (striped columns) is 100% in each region. The diagram has been constructed from the work of McGeer and McGeer (1976): value for AChE in CN is taken from Rinne *et al* (1973). Abbreviations: CN, caudate nucleus; GP, globus pallidus; SN, substantia nigra; ACB, nucleus accumbens; LC, locus coeruleus.

Stimulation of central dopamine systems in rodents results in hyperactivity, associated with low levels of stimulation, and various forms of stereotypy at high levels of stimulation. These so-called stereotyped behaviours are manifest as a wide range of activities including sniffing, licking, gnawing and head-bobbing. Conversely blockade of central dopamine mechanisms is associated with the state of catalepsy which, in rats, is regarded as the maintenance of an abnormal posture. Much research has tried, with partial success, to elucidate which anatomical substrates are associated with these various dopamine-dependent behaviours. Unfortunately the work is hampered by the lack of dopaminergic drugs which show specificity for either the extrapyramidal or limbic systems. The direct injection of dopamine itself or of putative dopamine agonists and antagonists has helped to elucidate the role of the various anatomical substrates in each of the behavioural responses, although the general conclusion must be a gross overlap between the functions of extrapyramidal and limbic dopamine structures.

Stimulation of one nigro-neostriatal dopamine complex in rodents results in the bizarre but striking phenomenon of circling behaviour (Anden *et al*, 1966). The behaviour is readily quantifiable and appears to be directly related to unilateral dopamine receptor stimulation (Glick *et al*, 1976). As such the rotating rodent is used commonly to investigate central dopamine mechanisms or central neurotransmitter interactions. The model is usually produced by the destruction of one nigro-neostriatal pathway either by an electrolytic lesion or by the injection of the neurotoxin 6-hydroxydopamine into the region of the dopamine cell bodies in the substantia nigra. Such an animal will rotate towards the lesioned side when given an indirectly-acting dopamine agonist such as amphetamine, due to release of endogenous dopamine from the terminals of the intact neostriatum, but away from the lesioned side when given a directly-acting agonist such as apomorphine, believed to be due to the preferential stimulation of the dopamine receptors on the lesioned side rendered supersensitive by degeneration of the dopaminergic terminals (Ungersted, 1971b). The general concepts of the rotating rodent have persisted for a decade, although new evidence suggests an involvement of limbic areas (notably the nucleus accumbens) as well as the extrapyramidal system (Kelly and Moore, 1976).

These various behavioural states in animals seen after manipulation of central dopamine function, together with their possible equivalent clinical situations, are summarised in Table I. One abnormality at

TABLE I
Manipulation of Central Dopamine Mechanisms

Dopamine Function	Anatomical Substrate	Animal Behaviour	Possible Clinical Situation
Stimulation (Dopamine agonists, e.g. apomorphine, amphetamine, piribedil bromocriptine)	Neostriatal and limbic structures	Stereotypy (repetitive behaviours)	Stereotypies (dyskinesias)
	Limbic structures	Hyperactivity	Psychotic states (dyskinesias)
Blockade (Dopamine antagonists, (neuroleptics), reserpine, a-methyl-p-tyrosine and lesions of dopaminergic pathways)	Neostriatal and limbic structures	Akinesia Hyperactivity	Akinesia Psychotic states (dyskinesias)
Unilateral stimulation	Neostriatum	Circling behaviour	Hemi-parkinsonism?

present appears to be the involuntary movement disorders (dyskinesias) The neuropharmacology of the dyskinesias is unclear, although indirect evidence suggests the possible overstimulation of certain populations of dopamine receptors (Klawans, 1973; Marsden, 1975). However in terms of animal (rodent) pharmacology, such behaviours described above provide reliable models with which to study possible functional interactions of other central neurotransmitters with the dopamine system.

Noradrenaline and Dopaminergic Function

The beneficial effects of L-DOPA for reversing the akinesia of Parkinson's disease are a clinical fact. L-DOPA (3, 4-dihydroxyphenylalanine) is the precursor substance of not only dopamine but also NA. It is therefore conceivable that the effects of L-DOPA in man and animals are the results of stimulation of both catecholamine mechanisms in the brain rather than those of dopamine alone. The results of animal experimental work suggest that NA does indeed have a role in certain aspects of L-DOPA induced behaviour.

The symptom of akinesia can be easily produced in mice by pretreatment with either reserpine, a drug known to deplete intraneuronal stores of monoamines (dopamine, NA and 5-HT), or α-methyl-p-tyrosine, a specific inhibitor of the catecholamine synthetic enzyme tyrosine hydroxylase. Normally such akinesia is immediately reversed by administration of L-DOPA, which has been shown to replenish central stores of dopamine and NA. However the complete reversal of the akinesia by L-DOPA can be prevented by pretreatment with agents which specifically block noradrenergic mechanisms (phenoxybenzamine, an α-adrenoceptor blocking agent; FLA 63 (bis [4-methyl-1-homopiperazinylthiocarbonyl] disulphide), an inhibitor of the NA synthesizing enzyme dopamine-β-hydroxylase) (Marsden *et al*, 1974; Dolphin *et al*, 1976). These results suggest that L-DOPA does indeed possess a noradrenergic component and that stimulation of both dopamine and NA receptors is required for the full restoration of motor activity in pharmacologically-induced akinetic animals. Additional support for this noradrenergic component is supplied by clonidine, a reported α-adrenoceptor agonist. Concurrent administration of clonidine greatly enhances the locomotor effects of dopamine agonists in rodents (Anden *et al*, 1973; Pycock *et al*, 1977).

This basic pharmacology implies a facilitatory action between the NA and dopamine systems of the brain. In turn it would suggest that the efficacy of L-DOPA treatment in Parkinson's disease may be related to the stimulation of central NA receptors as well as those of dopamine. The exact site of this facilitatory link between NA and dopamine is not known, but some experimental studies have suggested such a link may exist at the level of the substantia nigra. Circling behaviour usually occurs following a unilateral lesion of the nigro-neostriatal dopamine tract. However a circling rodent can be produced by lesioning one locus coeruleus (Pycock *et al*, 1975). This turning behaviour is still mediated through a dopamine mechanism and is not blocked by NA antagonists (Donaldson *et al*, 1976). Such an arrangement would indicate a facilitatory NA control over the nigro-neostriatal dopamine tract, probably at the level of the substantia nigra.

5-Hydroxytryptamine and Dopaminergic Function

Decreasing whole brain 5-HT concentrations following administration of parachlorophenylalanine (PCPA), an inhibitor of the synthesizing enzyme tryptophan hydroxylase, or lesioning the major 5-HT cell bodies in the raphe nuclei of the rostral brain stem is reported as increasing locomotor activity in rodents (Fibiger and Campbell, 1971; Jacobs *et al*, 1974). Conversely increasing cerebral 5-HT levels following administration of the precursors L-tryptophan or 5-hydroxytryptophan (5-HTP) is associated with decreased locomotor activity (Modigh, 1974). In general therefore facilitating central 5-HT mechanisms results in inhibition of the responses to dopamine receptor stimulation while blockade of 5-HT mechanisms enhances dopaminergic stimulation.

Bilateral injection of dopamine into the region of the nucleus accumbens results in a dose-dependent hyperactive response (Pijnenburg and Van Rossum, 1973). This behavioural response is markedly enhanced in animals with lesions of the raphe nuclei, or the response is depressed following the focal injection of 5-HT into the region of the nucleus accumbens (Costall *et al*, 1976). In a similar way injection of 5-HT bilaterally into the nucleus accumbens of rats blocks the increased locomotor response seen to the systemic administration of amphetamine (Fig. 2).

The state of catalepsy induced by dopamine receptor blockade with neuroleptic drugs can be enhanced by facilitating 5-HT transmission or reduced by blocking 5-HT mechanisms. For example haloperidol-induced catalepsy has been shown to be potentiated by systemic administration of 5-HTP or putative 5-HT agonists (quipazine, 5-methoxy-N, N-dimethyltryptamine) (Carter and Pycock, 1977). Conversely, neuroleptic catalepsy is decreased following

AMPHETAMINE HYPERACTIVITY & 5-HT INTO NUCLEUS ACCUMBENS

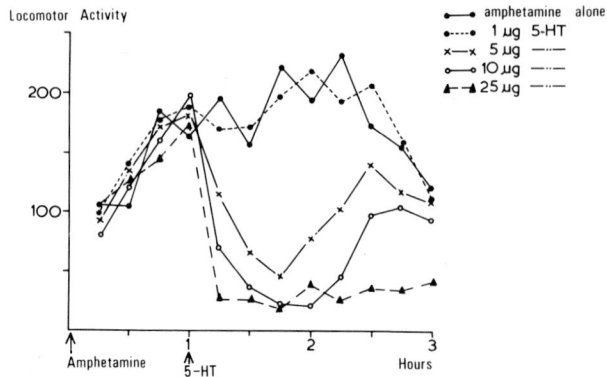

Fig. 2 Effect of bilateral injection of 5-HT (range 1−25 ug) into the nucleus accumbens on the locomotor response induced in rats by systemically administered amphetamine (1 mg/kg).

treatment with PCPA (Kostowski *et al*, 1972), the 5-HT antagonist cyproheptadine (Maj *et al*, 1975) and lesions of the raphe nucleus (Costall *et al*, 1975).

The results of these experiments suggest that 5-HT is inhibitory upon the dopamine system and perhaps a reciprocal relationship exists between 5-HT and dopamine in the brain. This hypothesis is borne out in experiments showing that elevating 5-HT transmission will inhibit circling behaviour in mice induced by dopamine agonists while blockade of 5-HT mechanisms enhances dopamine-dependent circling (Milson and Pycock, 1975).

Acetylcholine and Dopaminergic Function

The beneficial use of anticholinergic drugs in the treatment of Parkinson's disease supports the hypothesis of a reciprocal balance between dopamine and ACh in the striatum (Barbeau, 1962). The current general opinion is that the dopaminergic input into the striatum normally inhibits striatal cholinergic neurones (Bartholini *et al*, 1973). Loss of the dopaminergic input in Parkinson's disease produces an overactivity in the striatal ACh system which is responsible for some of the characteristic motor disturbances such as tremor (Hornykiewicz, 1971).

This possible synaptic arrangement receives support from behavioural studies. For example anticholinergic drugs induce hyperactivity (Sanger and Steinberg, 1974) and potentiate the behavioural effects of dopamine agonists (Arnfred and Randrup, 1968). Conversely cholinergic drugs produce catalepsy in rodents (Ahtee and Kaariainen, 1974) and block dopamine-dependent circling activity (Marsden *et al*, 1975). Interpretation of these dopamine: ACh behavioural data should also take into account the ACh innervation identified in the region of the substantia nigra (Butcher *et al*, 1975) which may be inhibitory on the nigro-neostriatal dopaminergic activity. However in the parkinsonian state where the dopamine tract is degenerating the cholinergic influence in the substantia nigra will probably be minimal.

GABA and Dopaminergic Function

GABA is postulated as an inhibitory neurotransmitter in the brain (Roberts, 1974). Although GABA has a ubiquitous distribution and its neuroanatomical connections in the brain are not clear, a GABA-mediated pathway from the striopallidal complex to the substantia nigra has been proposed (see McGeer, McGeer and Hattori, 1977). Such a pathway is believed to exert an inhibitory control on the nigro-neostriatal dopaminergic tract, and this together with the fact that GABA is found in high concentrations in the dopamine-containing striatal and limbic regions suggests a functional inter-relationship between these two transmitter systems.

Elevation of cerebral GABA concentrations is associated with akinesia, whilst blockade of central GABA function evokes hyper-activity, myoclonic jerks and generalized seizures (Pycock *et al*, 1976). In keeping with these concepts the focal injection of GABA bilaterally into the nucleus accumbens inhibits both the hyperactivity induced by injection of dopamine into this same site and the increased locomotor effects of systemically administered amphetamine (Fig. 3). The effects are dose-dependent but only last approximately 30 minutes, probably due to efficient GABA inactivation mechanisms (uptake). In addition drugs which are believed to stimulate central GABA mechanisms have been shown to potentiate the cataleptic response seen after dopamine receptor blockade (Kaarianinen, 1976).

Manipulation of GABA systems in the region of one substantia nigra in rats evokes circling behaviour. For example increasing GABA concentrations unilaterally creates an animal which will rotate towards the side of the higher GABA concentration in response to systemically applied dopamine agonists (Dray *et al*, 1975), while injection of the GABA receptor blocking agent picrotoxin into one substantia nigra causes circling away from the injected side (Tarsy *et al*, 1975; Pycock, 1976). Such results are in keeping with an inhibitory GABA input regulating the activity of the nigro-neostriatal

GABA & LOCOMOTOR ACTIVITY

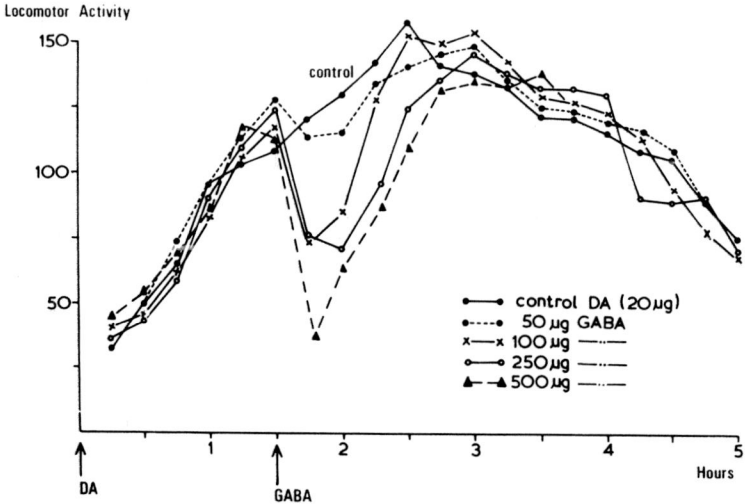

A) Dopamine Hyperactivity from N. Accumbens

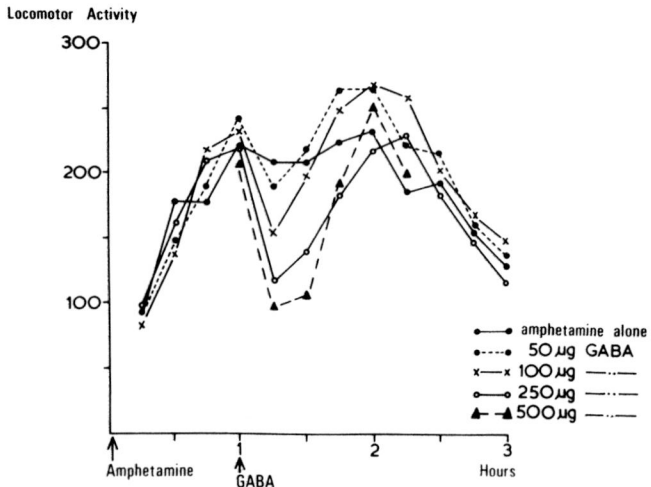

B) Amphetamine Hyperactivity

Fig. 3 Effect of bilateral injection of GABA (range 50–500 ug) into the nucleus accumbens on a) hyperactivity response induced by bilateral injection of dopamine (20 ug) into the same site 90 min previously in nialamide-pretreated rats (50 mg/kg, i.p., 2 hr before dopamine) and b) locomotor response induced in rats by systemically administered amphetamine (1 mg/kg).

dopamine pathway.

Amino Acid and Peptide Neurotransmitter Candidates and Dopaminergic Function

To date the interaction of only a few central transmitter candidates with dopamine mechanisms has been investigated. The possible importance of other amino acid neurotransmitters and their role in neurological disorders like Parkinson's disease must not be overlooked. For example there has been the suggestion of a cortico-striatal projection whose transmitter may be the excitatory amino acid glutamic acid (Spencer, 1976; McGeer, McGeer, Scherer and Singh, 1977). What role such a projection might play in normal basal ganglia function is as yet unknown. Taurine is another amino acid found in appreciable quantities within the basal ganglia. Although in preliminary observations no changes in taurine concentrations were detected in parkinsonian brains (Perry, 1976) the possible interactions of this putative central transmitter with forebrain dopamine mechanisms are not yet known.

Among the putative peptide neurotransmitters in the brain, the undecapeptide substance P has received much attention. Substance P is widely but unevenly distributed in the CNS, with high concentrations in the substantia nigra (Kanazawa and Jessell, 1976). Electrophysiological studies suggest an excitatory role for substance P in the substantia nigra (Davies and Dray, 1976) and behavioural work following intranigral injections of substance P supports an excitatory action on the nigro-neostriatal dopamine tract (James and Starr, 1977; Olpe and Koella, 1977). The functional role of nigral substance P

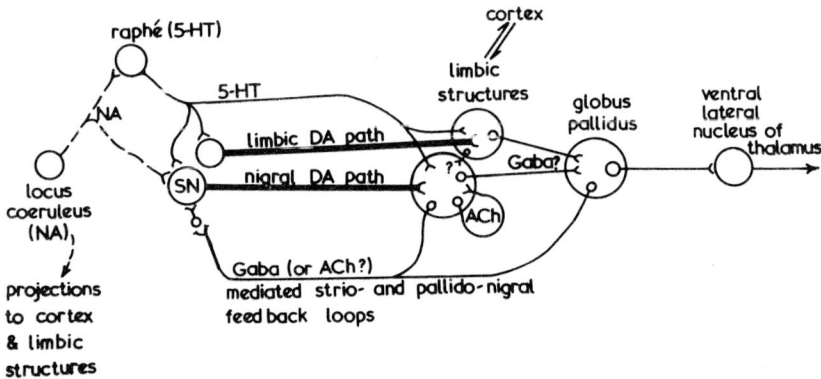

Fig. 4 Some major neuronal connections of the dopamine (DA) pathways in the basal ganglia. (For possible involvement of other amino acids, peptides or opiate interactions see text.) SN denotes substantia nigra.

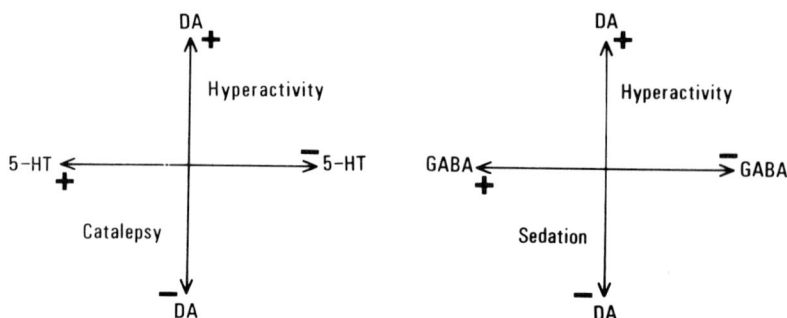

Fig. 5 Hypothetical representation of the interaction between 5-HT and dopamine (DA) and GABA and dopamine in the behavioural states of hyperactivity and catalepsy or sedation.

has yet to be established although there is evidence for the existence of substance P-containing fibres from the striato-pallidal complex to the substantia nigra, suggesting a feedback role to control activity of the nigro-neostriatal dopamine tract (Hong *et al*, 1977; Kanazawa *et al*, 1977). Such a regulatory system might imply an important role for substance P in neurological disorders such as Parkinson's disease although as yet there are no published data on this peptide in parkinsonian brains.

A further consideration on transmitter interactions and the dopamine system is the world of opiate neurochemistry. Certain opiate agonists (endorphins) have been observed to bind in many regions of the brain. (Endorphins are peptides of which enkephalin is a specific endogenous morphine-like factor.) As with the opiate receptor, enkephalin activity is associated with limbic and extra-pyramidal areas of the brain (Simatov *et al*, 1976). β-Endorphins can induce catatonic states in rats reminiscent of some aspects of schizophrenia (Bloom *et al*, 1976). The role of these and other peptides in neurological disorders has yet to be assessed.

Functional Relationships between Dopamine and Other Neurotransmitters

The above experimental data provide evidence for a functional inter-action between dopamine and other neurotransmitters in the basal ganglia and limbic regions of the brain. The current neuronal systems and their sites of possible functional interaction with the dopamine pathways are illustrated in Fig. 4. Apart from the facilitatory effects of NA stimulation, many of the interactions appear to be inhibitory, and as such we must predict a reciprocal balance for 5-HT, ACh and GABA with the dopamine system. Such a relationship for 5-HT and

TABLE II

Other Neurotransmitters in Parkinson's Disease: Relationship to Dopamine

Neurotransmitter	Overall Effect on Dopamine Function	Drugs of Theoretical Value in Parkinson's Disease
NORADRENALINE	facilitatory	NA agonists e.g. L-DOPA clonidine uptake blockers MAO inhibitors
5-HYDROXYTRYPTAMINE	inhibitory	5-HT receptor blockers e.g. methysergide cyproheptadine metergoline
GABA	inhibitory	GABA receptor blockers e.g. picrotoxin bicuculline GAD inhibitors e.g. allylglycine
ACETYLCHOLINE	inhibitory	Anticholinergic drugs e.g. benztropine benzhexol
AMINO ACIDS (glutamic acid, taurine)	+ or −	?
PEPTIDES (substance P, enkephalin and other endorphins)	+ or −	? (opiate neurochemistry)

GABA with dopamine is illustrated diagrammatically in Fig. 5, which hypothesises this reciprocal correlation between the transmitters to explain the behavioural patterns observed. Hyperactivity and catalepsy (sedation) represent opposite poles of the spectrum of dopamine activity, and their clinical equivalents may be Parkinson's disease at one end and dyskinesias and psychoses at the other. The relationship of other central neurotransmitters to dopamine function is summarised in Table II, together with a list of agents which, based on animal studies, are of theoretical use in the treatment of Parkinson's disease.

Summary and Conclusions

While the bulk of pathological and neurochemical evidence focuses on the loss of the nigro-neostriatal dopamine system in Parkinson's disease, supplementary data reveal the additional loss, and thus the possible implication, of other central neurotransmitter substances in this disorder. However at this time it is not possible to separate the pathology which is a secondary result of the disease from that which is the primary cause of the disorder. Certainly examination of parkinsonian brains reveals lesions in the striatum, globus pallidus and brain stem as well as in the substantia nigra, while neuro-chemistry records gross reductions in the concentrations of other putative neurotransmitter substances (NA, 5-HT, GABA) and perhaps more importantly of their synthetic enzymes. Further consider-ations of these aspects of Parkinson's disease may provide new therapeutic approaches for its clinical management.

Acknowledgement

Financial support from the Parkinson's Disease Society and the National Fund for Crippling Disease is gratefully acknowledged. I wish to thank Mrs. Pat Berman and Mrs. Ann Duncan for help in the preparation of this manuscript.

References

Ahtee, L. and Kaariainen, I. (1974). *N. S. Arch. Pharmacol.* **284**, 25—38.
Anden, N.-E., Dahlstrom, A., Fuxe, K. and Larsson, K. (1966). *Acta Pharmacol. Toxicol.* **24**, 263—274.
Anden, N.-E., Strombom, U. and Svensson, T.H. (1973). *Psychopharmacologia (Berl.)* **29**, 289—298.
Arnfred, T. and Randrup, A. (1968). *Acta Pharmacol. Toxicol.* **26**, 384—394.
Barbeau, A. (1962). *Canada. Med. Ass. J.* **87**, 802—807.
Bartholini, G., Stadler, H. and Lloyd, K.G. (1973). *Adv. Neurol.* **3**, 233—241.
Bernheimer, H., Birkmayer, W. and Hornykiewicz, O. (1963). *Klin. Wschr.* **41**, 564—569.
Bird, E.D. and Iversen, L.L. (1974). *Brain* **97**, 457—472.
Bloom, F., Segal, D., Ling, N. and Guillemin, R. (1976). *Science* **194**, 630—632.
Butcher, L.L., Talbot, K. and Bilezikjian, L. (1975). *J. Neural Trans.* **37**, 127—153.
Carter, C.J. and Pycock, C.J. (1977). *Brit. J. Pharmacol.* **60**, 267—268P.
Costall, B., Fortune, D.H., Naylor, R.J., Marsden, C.D. and Pycock, C. (1975). *Neuropharmacol.* **14**, 859—868.
Costall, B., Naylor, R.J., Marsden, C.D. and Pycock, C.J. (1976). *J. Pharm. Pharmacol.* **28**, 523—526.
Davies, J. and Dray, A. (1976). *Brain Res.* **107**, 623—627.
Dolphin, A.C., Jenner, P. and Marsden, C.D. (1976). *Pharmacol. Biochem. Behav.* **4**, 661—670.
Donaldson, I.McG., Dolphin, A., Jenner, P., Marsden, C.D. and Pycock, C. (1976). *Eur. J. Pharmacol.* **39**, 179—191.
Dray, A., Oakley, N.R. and Simmonds, M.A. (1975). *J. Pharm. Pharmacol.* **27**, 627—629.
Farley, I.J. and Hornykiewicz, O. (1976). *In* "Advances in Parkinsonism,

Biochemistry, Physiology, Treatment" (eds. Birkmayer, W. and Hornykiewicz, O.), Editiones Roche, Basel.

Farley, I.J., Price, K.S. and Hornykiewicz, O. (1977). *In* "Advances in Biochemical Psychopharmacology" (eds. Costa, E. and Gessa, G.L.), Vol. 16, pp. 57–64. Raven Press, New York.

Fibiger, H.C. and Campbell, B.A. (1971). *Neuropharmacol.* **10**, 25–32.

Glick, S.D., Jerussi, T.P. and Fleisher, L.N. (1976). *Life Sci.* **18**, 889–896.

Hong, J.S., Yang, H.-Y.T., Racagni, G. and Costa, E. (1977). *Brain Res.* **122**, 541–544.

Hornykiewicz, O. (1966). *Pharmacol. Rev.* **18**, 925–964.

Hornykiewicz, O. (1971). *In* "Recent Advances in Parkinson's Disease" (eds. McDowell, F.H. and Markham, C.H.), pp. 33–65. Davis, Philadelphia, Pa.

Hornykiewicz, O. (1972). *In* "Handbook of Neurochemistry" (ed. Lajtha, A.), Vol. 7, pp. 465–501. Plenum Press, New York.

Hornykiewicz, O., Lloyd, K.G. and Davidson, L. (1976). *In* "GABA in Nervous System Function" (eds. Roberts, E., Chase, T.N. and Tower, D.B.), pp. 479–485. Raven Press, New York.

Jacobs, B.L., Wise, W.D. and Taylor, K.M. (1974). *Brain Res.* **79**, 353–361.

James, T.A. and Starr, M.S. (1977). *J. Pharm. Pharmacol.* **29**, 181–182.

Kaariainen, I. (1976). *Acta Pharmacol. Toxicol.* **39**, 393–400.

Kanazawa, I., Emson, P.C. and Cuello, A.C. (1977). *Brain Res.* **119**, 447–453.

Kanazawa, I. and Jessell, T. (1976). *Brain Res.* **117**, 362–367.

Kelly, P.H. and Moore, K.E. (1976). *Nature (Lond.)* **263**, 695–696.

Klawans, H.L. (1973). *In* "The Pharmacology of Extrapyramidal Movement Disorders" (ed. Cohen, M.M.), Monographs in Neural Sciences, Vol. 3, pp. 1–136. Karger, Basle.

Kostowski, W., Gumulka, W. and Czlonkowski, A. (1972). *Brain Res.* **48**, 443–446.

Lindvall, O. and Bjorklund, A. (1974). *Acta Physiol. Scand., Suppl.* **412**, 1–48.

McGeer, P.L. and McGeer, E.G. (1976). *J. Neurochem.* **26**, 65–76.

McGeer, P.L., McGeer, E.G. and Hattori, T. (1977). *In* "Advances in Biochemical Psychopharmacology" (eds. Costa, E. and Gessa, G.L.), Vol. 16, pp. 397–402. Raven Press, New York.

McGeer, P.L., McGeer, E.G., Scherer, U. and Singh, K. (1977). *Brain Res.* **128**, 369–373.

Maj, J., Mogilnicka, E. and Przewlocka, B. (1975). *Pharmacol. Biochem. Behav.* **3**, 25–27.

Marsden, C.D. (1975). *In* "Modern Trends in Neurology" (ed. Williams, D.), Vol. 6, pp. 141–166. Butterworths, London.

Marsden, C.D., Dolphin, A., Duvoisin, R.C., Jenner, P. and Tarsy, D. (1974). *Brain Res.* **77**, 521–525.

Marsden, C.D., Milson, J., Parkes, J.D., Pycock, C. and Tarsy, D. (1975). *J. Physiol. (Lond.)* **249**, 64–65P.

Milson, J.A. and Pycock, C.J. (1975). *Brit. J. Pharmac.* **56**, 77–85.

Modigh, K. (1974). *Acta Physiol. Scand., Suppl.* **403**, 1–56.

Olpe, H.-R. and Koella, W.P. (1977). *Brain Res.* **126**, 576–579.

Perry, T.L. (1976). *In* "Taurine" (eds. Huxtable, R. and Barbeau, A.), pp. 365–374. Raven Press, New York.

Pijnenburg, A.J.J. and Van Rossum, J.M. (1973). *J. Pharm. Pharmacol.* **25**, 1003–1005.

Pycock, C.J. (1976). *In* "Biochemistry and Neurology" (eds. Bradford, H.F. and Marsden, C.D.), pp. 93–102. Academic Press, London.

Pycock, C.J., Donaldson, I.McG. and Marsden, C.D. (1975). *Brain Res.* **97**, 317–329.

Pycock, C., Horton, R.W. and Marsden, C.D. (1976). *Brain Res.* **116**, 353–359.

Pycock, C.J., Jenner, P.G. and Marsden, C.D. (1977). *N.S. Arch. Pharmacol.* **297**, 133–141.

Rinne, U.K., Riekkinen, P., Sonninen, V. and Laaksonen, H. (1973). *Acta Neurol. Scand.* **49**, 215–226.

Roberts, E. (1974). *Biochem. Pharmacol.* **23**, 2637–2649.

Sanger, D.J. and Steinberg, H. (1974). *Eur. J. Pharmacol.* **28**, 344–349.

Simantov, R., Kuhar, M.J., Pasternak, G.W. and Snyder, S.H. (1976). *Brain Res.* **106**, 189–197.

Spencer, H.J. (1976). *Brain Res.* **102**, 91–101.

Tarsy, D., Pycock, C., Meldrum, B. and Marsden, C.D. (1975). *Brain Res.* **89**, 160–165.

Ungerstedt, U. (1971a). *Acta Physiol. Scand., Suppl.* **367**, 1–48.

Ungerstedt, U. (1971b). *Acta Physiol. Scand., Suppl.* **367**, 69–93.

MOVEMENT, GROWTH HORMONE AND PROLACTIN RESPONSES TO DOPAMINE AND DOPAMINE AGONISTS

J.D. PARKES, P.A. PRICE and C.D. MARSDEN

University Department of Neurology,
King's College Hospital and the Institute of Psychiatry, London, U.K.

Introduction

Dopamine is involved in the function of the nigrostriatal, growth hormone and prolactin systems, and in this brief review we will consider some aspects of the timing, magnitude, and variability of the movement and hormonal responses to dopamine. Much of the information has been gathered from studies of patients with Parkinson's disease, acromegaly or hyperprolactinaemia, and it must be remembered that the responses to dopamine in these conditions may be abnormal in the presence of damage to nigrostriatal or hypothalamic tubero-infundibular dopamine systems. Also many hormonal responses are influenced by stress and vomiting, and since dopamine is involved in the physiologic control of blood pressure, renal blood flow, autonomic control and many other systems, change in the activity of these variables may affect hormonal responses. A recent summary of the extracerebral distribution of dopamine systems has been given by Thorner (1975).

The movement response to dopamine stimulants in parkinsonism may depend upon neuronal circuits with a dopamine-neurone/amine-neurone link; the growth hormone response upon a dopamine-neurone/peptide-neurone connection; and the prolactin response upon a dopamine-neurone/portal circulation/pituitary circuit. In the first two instances, dopamine would function as a conventional neuro-transmitter; in the last instance, as a local hormone. In each case, the final response is likely to depend upon the activation of dopamine receptors on cell surfaces. There has been considerable speculation as to the possible existence of different types of dopamine receptors, but there is no present evidence from the action of dopamine stimulants or antagonists in man for the presence of radically different dopamine receptor types in movement, growth hormone and prolactin systems, although, as will be discussed, there may be subdivisions in dopamine receptor types mediating movement.

The main factors determining response to dopamine stimulants include variation in absorption and metabolism, difference in penetration into different areas both inside and outside the brain, and different drug actions on the various neurotransmitter systems in the brain. Thus, the absorption of levodopa is rapid, peak levels in the plasma are achieved within 60—90 minutes of dosage, and the drug is rapidly cleared from the plasma. In contrast, peak plasma bromocriptine levels are achieved a little later, and the drug can still be determined in the plasma 24 hours after a single dose (Fig. 1). The actions of levodopa are largely attributable to enzymic conversion to dopamine, and possibly noradrenaline; whilst in the case of bromocriptine various metabolites may be the active compounds.

Bromocriptine has a weak antagonist effect on noradrenaline systems. Dopamine is manufactured and excreted by the kidney, whereas one main avenue of bromocriptine excretion is biliary. In addition to these differences in pharmacokinetics, it is possible that different kinds of dopamine receptor are present in the central nervous system, analagous to α and β adrenergic receptors, and that drugs may have type-specific receptor activity. However, there is at present no definite evidence for this view in man. Yet another factor determining response to dopamine stimulants and antagonists is the possible development of receptor supersensitivity, during denervation or with long term treatment (Pycock and Marsden, 1976). All these factors are likely to be involved in the clinical responses to levodopa and bromocriptine in the treatment of parkinsonism, acromegaly and hyperprolactinaemia.

Dopamine and Parkinson's Disease

No satisfactory explanation has yet been given for the phenomenon of delayed clinical response to levodopa during the initial period of treatment. However, when levodopa is combined with a decarboxylase inhibitor, improvement is often obvious within the first week, and once a patient is established on treatment, there is frequently a definite increase in mobility 1—2 hours after each separate dose and a decrease in mobility after 3—4 hours. This period of drug effectiveness almost certainly decreases with continued treatment, or alternatively with the progression of disease. This results in the gradual appearance of a highly unstable response to multiple oral doses of levodopa, although during the first 1—2 years of levodopa treatment diurnal response variations are usually slight. The addition of decarboxylase inhibitors does not affect this pattern of response. In

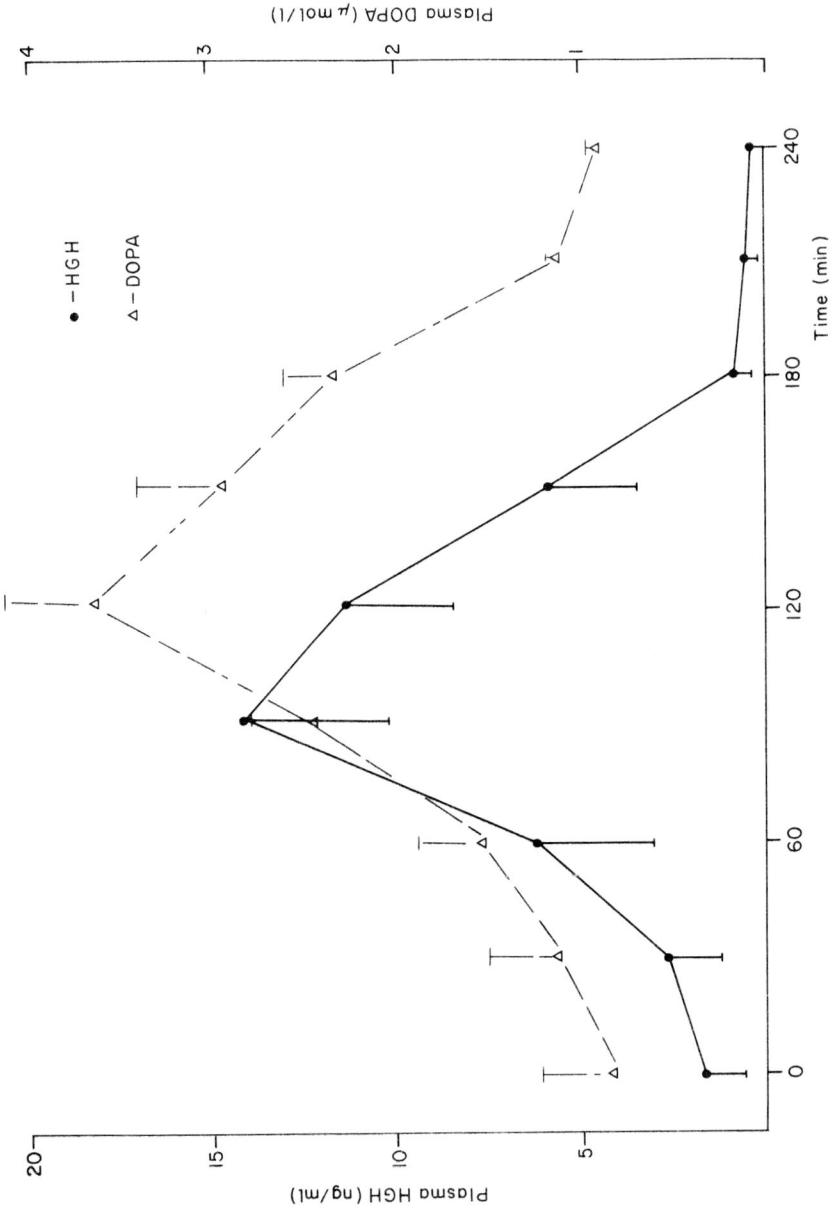

Fig. 1a. Plasma DOPA and Bromocriptine levels in Parkinsonian subjects following: Levodopa 1 g po.

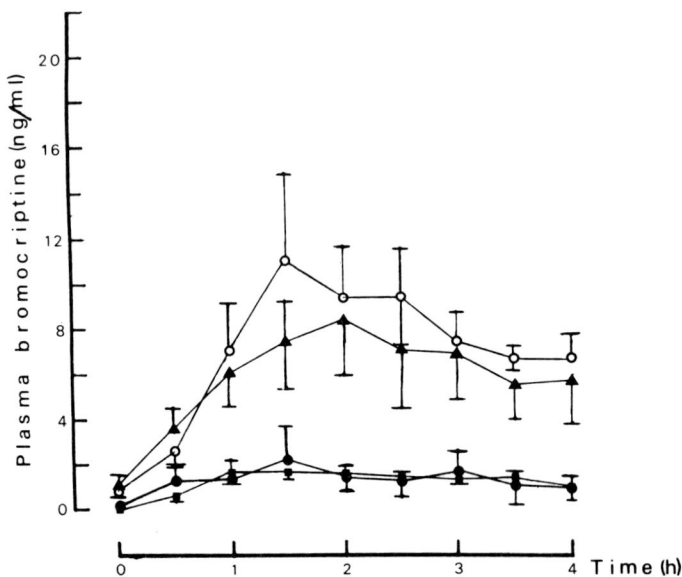

Fig. 1b. Plasma DOPA and Bromocriptine levels in Parkinsonian subjects following: Bromocriptine 12.5mg po (■); Bromocriptine 25mg (●); Bromocriptine 50mg (▲); and Bromocriptine 100mg (○). Mean plasma levels $\pm$. 1 SEM are shown. The plasma growth hormone response to Levodopa 1g is illustrated. Figure and data on plasma Bromocriptine levels reproduced with permission from Price *et al*, British Journal of Clinical Pharmacology; in press.

theory, and possibly also in practice, bromocriptine gives a steadier response.

There are many different kinds of response fluctuations in parkinsonism as well as those related in time to separate oral doses of levodopa or bromocriptine (Marsden and Parkes, 1976). These include early morning akinesia, freezing episodes, disability due to severe involuntary movements, and possibly akinesia accompanying extremely high plasma dopa levels, so-called peak dose akinesia. The commonest type of response fluctuation in untreated patients is due to freezing, in which the patient confronted with a difficult motor task suddenly becomes unable to move. At other times, and classically in dangerous situations, near-normal movement may be restored for a brief period. Freezing mainly occurs in severely disabled subjects, and when it is superimposed on response swings due to rapid fluctuations in plasma dopa levels in treated patients, it causes severe and often unpredictable fluctuations in mobility. Response swings with changing plasma dopa levels can be avoided by maintaining steady plasma levels with intravenous levodopa (Shoulson *et al*, 1975), but other factors as well as the availability of dopa to the brain must be involved in the severe response swings of some patients. These factors probably include changes in the concentration of other amino acids, changes in activity in non-dopaminergic neuro-transmitter systems, and alterations in the concentrations of brain peptides and growth hormone.

Competition between different amino acids and dopa for penetration into the brain occurs in animals, but drastic reduction of dietary protein intake rarely alters the degree of stability of levodopa control in parkinsonism. Cotzias *et al* (1976) suggested the hypothesis that changes in growth hormone concentration as a result of levodopa could alter the central metabolism and response to dopamine, and hence result in changes in clinical response. However, there seems to be little relationship between clinical response, dyskinesias, and growth hormone levels in parkinsonism (Debono *et al*, 1977). The development of response swings in patients with parkinsonism on levodopa is likely to be accompanied by a progressive loss of substantia nigra neurones, and possibly loss of a sustained capacity to manufacture dopamine from levodopa.

Dyskinesias in Parkinsonism

Costall and Naylor in this symposium (Chap. 9, p. 129) have presented evidence for the presence of specific 'dyskinesia receptors' in animals and suggested that it may be possible to block dyskinesias, but not the other movement responses to dopamine stimulants in parkinsonism, by drugs such as oxiperomide and tiapride. There are certainly many bizarre clinical observations of dyskinesias that may point to the existence of a population of dopamine-sensitive dyskinesia receptors of a different type from other dopamine receptors. Cools and van Rossum (1976) have suggested the presence of two distinct dopamine receptor types, signified as DAi and DAe respectively (i indicating inhibition by dopamine, and e, excitation) and detailed the following clinical observations:

a) In Huntington's chorea and parkinsonism, levodopa causes chorea whilst the dopamine agonist apomorphine may apparently relieve chorea.

b) Both dopamine stimulants and dopamine antagonists, used separately, are occasionally successful in the treatment of torsion dystonia.

c) The combination of a dopamine-releasing drug, amantadine, with a receptor blocking drug, haloperidol, may be of value in the treatment of spasmodic torticollis.

d) The long term treatment of parkinsonism with levodopa, and of psychotic patients with phenothiazines, both result in dyskinesias.

e) Piribedil, a dopamine stimulant that is highly effective in animal models of parkinsonism, is not very active in man.

f) Levodopa affects the motor behaviour of people with parkinsonism, but not that of normal subjects.

None of this apparently conflicting data gives evidence for the presence of distinct dopamine receptor types in man. Dyskinesias are notoriously variable and clinical observation of them gives conflicting results. Apomorphine metabolites may possess dopamine antagonist actions, and in animals, the stimulant action of piribedil is dependent upon the formation of an active metabolite. A dopaminergic defect has not convincingly been shown in torsion dystonia.

At present, our clinical studies suggest that chlorpromazine 50–250mg, pimozide 8–16mg, and to a lesser extent metoclopramide 20–60mg, will partially or completely abolish dyskinesias resulting from levodopa 1g, but only at the expense of some reduction in clinical response. However, it seems probable that oxiperomide may prevent dyskinesias without altering the clinical response (Donaldson, personal communication). If this is so, it offers an exciting prospect of eventually separating some of the various actions of different dopamine agonists.

Growth Hormone and Dopamine

Levodopa causes a rise in plasma growth hormone level in normal subjects, but a fall in (the elevated) level in acromegalic subjects. The rise in normal subjects is prevented by chlorpromazine and other drugs causing dopamine receptor blockade. Other dopamine stimulants also cause an increase in plasma growth hormone levels in non-acromegalic subjects. Noradrenaline, as well as dopamine, influences plasma growth hormone levels in man, but the effect of α and β receptor stimulation or blockade is comparatively slight compared with the action of dopamine.

Following an oral dose of levodopa in normal subjects, the time-curve of growth hormone response corresponds closely to elevation of plasma dopa levels, with an initial growth hormone response at 30–60 minutes and peak levels at 90–120 minutes (Fig. 2). Growth hormone levels fall to fasting values after 3–4 hours. A similar growth hormone plasma time-curve is seen after bromocriptine, despite prolongation of plasma bromocriptine levels for many hours. The time course of elevation of plasma growth hormone levels by dopamine stimulants otherwise corresponds fairly well with the reported time courses of activation of dopamine receptors, as indicated by change

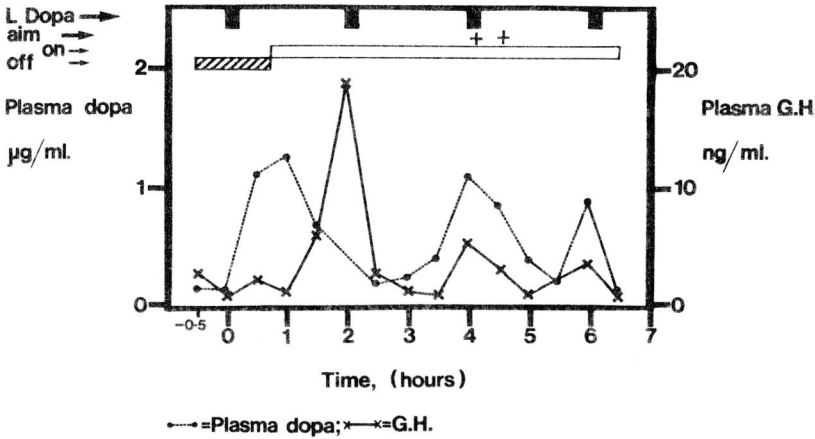

Fig. 2 Variation in plasma DOPA level and accompanying changes in the clinical and growth hormone response in a single patient with Parkinsonism given four separate oral doses of Levodopa 1g at two hourly intervals.
Aim indicates the presence of dyskinesias, 'on' indicates periods of mobility and 'off' indicates periods of immobility.

in behaviour, such as mobility, reversal of neuroleptic-catalepsy, and circling. On the whole, maximum improvement in parkinsonism, dyskinesias, and peak plasma growth hormone levels in response to levodopa occur simultaneously, and at approximately the same time as peak plasma dopa levels.

Approximately 5–10% of normal and parkinsonian subjects do not have an obvious increase in plasma growth hormone level following levodopa 1g. The rate of change, rather than absolute values, of dopamine concentration achieved in the brain may be a major factor in determining whether a levodopa-growth hormone response occurs or not; this appears to be the case in respect of blood sugar concentration in determining whether an insulin hypoglycaemic-growth hormone response occurs or not. However, in the case of levodopa, a minimal basal plasma dopa level of 400ug/ml may be necessary for an unequivocal plasma growth hormone response to occur (Mars and Genuth, 1973).

In normal subjects, the growth hormone response to levodopa is likely to be due to a hypothalamic action of dopamine, resulting in release of the peptide growth hormone releasing factor or inhibition of growth hormone release inhibiting factor (somatostatin). Somewhat surprisingly, somatostatin has widespread effects and causes a decrease in pancreatic insulin and glucagon output. If

dopamine is inhibitory to the production of somatostatin at a hypothalamic level, this may account for the reported increase of plasma
glucagon levels following levodopa (Rayfield *et al*, 1975). However,
somatostatin has an extremely short half life in the circulation,
measured in minutes only.

Some of the evidence for a hypothalamic action of levodopa and
hence dopamine in causing a rise in plasma growth hormone levels is
summarised in Fig. 3 (from Price and Parkes, unpublished data). The
anatomical basis for growth hormone control by dopamine is shown
by the presence of dopamine neurones in the tubero-infundibular
regions.

In the 1960s, it became apparent that catecholamines affect
pituitary growth hormone content in animals. Dopamine added to
cultures of normal animal or human pituitaries *in vitro* has no action,
but it depletes growth hormone stores *in vivo*. Direct intrahypothalamic
injections of dopamine result in an elevation of plasma growth hormone
levels, and this effect is abolished by pituitary stalk section. Elevation
of plasma growth hormone levels following dopamine can be prevented
by catecholamine receptor blockade (chlorpromazine, pimozide) or

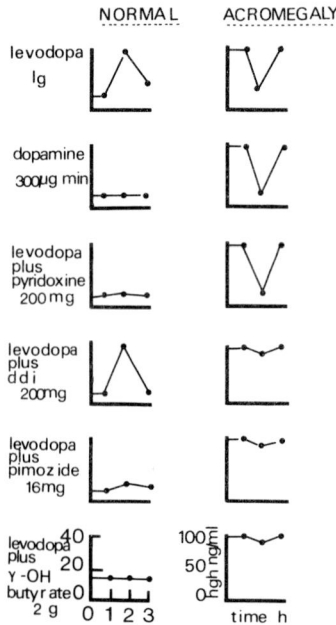

Fig. 3   Diagramatic representation of the growth hormone response to different drugs and
drug combinations in normal and acromegalic subjects.
'ddi' indicates the decarboxylase inhibitor, L-alphamethyl DOPA hydrazine.

by prevention of catecholamine synthesis or storage (p-amino-methyl-tyrosine, reserpine, see Martin, 1973, for review). Pyridoxine, which increases the peripheral metabolism of levodopa, abolishes the growth hormone response to levodopa in man, and decarboxylase inhibitors, which do not penetrate the brain, enhance the response. Gamma hydroxybutyrate, which prevents the release of dopamine from dopamine neurones, also abolishes the levodopa response.

Chlorpromazine abolishes the growth hormone response to dopamine in animals, but not the growth hormone response to hypothalamic extracts containing growth hormone releasing factors (Muller *et al*, 1967). It seems probable therefore that dopamine causes the release of growth hormone releasing factor in the hypothalamus, and this peptide then enters the portal circulation to reach the pituitary. In contrast to this peptide link in dopamine-growth hormone control, there is no evidence of a peptide link in dopamine-nigral control of movement. (Substance P occurs in high concentration in the substantia nigra, but there is no evidence that it is directly released in response to dopamine, and it is found mainly in cholinergic terminals.) Such a peptide link in growth hormone control may account for a slight delay in the growth hormone response to elevation of plasma dopa in normal subjects. However, the exact details of response timings in the minutes following dopa infusion are not known in either normal or acromegalic subjects.

There is some evidence that the apparently paradoxical effect of dopamine stimulants in acromegalic subjects is due to an alteration in membrane characteristics of the acromegalic somatotroph, which becomes directly sensitive to, and inhibited by, dopamine (Verde *et al*, 1976). Chlorpromazine causes a fall in plasma growth hormone levels in both normal and acromegalic subjects (Kolodny *et al*, 1971), although metoclopramide may partially prevent bromocriptine suppression of growth hormone levels in acromegalics (Delitala *et al*, 1976). As discussed, the reduction in plasma growth hormone level following levodopa in acromegalics is rapid in onset, and following bromocriptine, is sustained for several hours. The duration of growth hormone response to bromocriptine in acromegalics corresponds closely to the duration of motor hyperactivity in animals, although motor excitation is delayed rather than immediate in onset (Johnson *et al*, 1976).

The steady prolonged suppression of growth hormone levels by bromocriptine in acromegalics continues for several years of treatment (Thorner and Besser, 1976), and does not come to resemble the brittle 'on-off' response to levodopa and, to a lesser extent, to

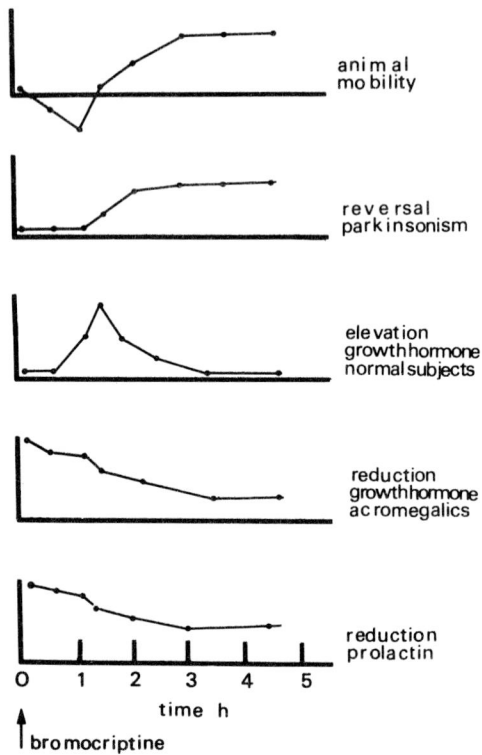

Fig. 4 Diagramatic representation of behavioural and hormonal response in animals and in man to a single dose of Bromocriptine po.

bromocriptine, seen in parkinsonism.*

Many dopamine stimulants, including bromocriptine, cause an initial phase of locomotor inhibition in animals before a more prolonged period of motor excitation occurs (Fig. 4). Whilst the cause of motor inhibition is not known an attractive hypothesis is the idea that low concentrations of dopamine stimulants may preferentially stimulate pre-synaptic, and high concentrations, post-synaptic receptors. Pre-synaptic dopamine autoreceptor stimulation may result in a reduction, and not an increase, of pre-synaptic dopamine synthesis and release. In favour of this idea is the observation that parkinsonian disability occasionally increases rather than decreases

*However, following prolonged bromocriptine treatment in acromegalics, it has recently been reported that an elevation and not a depression of plasma growth hormone levels results from levodopa.

during the initial period of bromocriptine treatment, but we have not observed this effect. It is not at present known whether low doses of levodopa, apomorphine or bromocriptine have opposite effects to high doses on the growth hormone response in either normal or acromegalic subjects.

Prolactin and Dopamine

Levodopa and dopamine stimulants cause a reduction in plasma prolactin levels in normal subjects, in those with Parkinson's disease, and in patients with hyperprolactinaemia. Dopamine is one of the major physiologically active factors that cause suppression of pituitary prolactin output and suppress prolactin levels in the blood (Schally *et al*, 1974). There are, however, other prolactin inhibitory factors, as shown by the findings that levodopa causes only a 50% reduction in prolactin levels in normal subjects; that hypothalamic extracts with no catecholamine content will suppress prolactin levels; and also that in hyperprolactinaemic subjects, levodopa is not as effective as hypophysectomy in depressing prolactin levels. In contrast to levodopa and dopamine, the decarboxylase inhibitor 1-α-methyl dopa hydrazine causes an increase in pituitary prolactin output, but given without levodopa it has no effect on plasma growth hormone levels or on parkinsonian symptoms.

The evidence that dopamine suppresses pituitary prolactin production was gained in the 1970's following the development of specific radioimmunoassay techniques for the determination of human prolactin. Unlike the situation with growth hormone, dopamine suppresses the output of prolactin from the pituitary both *in vitro* and *in vivo* (Pasteels *et al*, 1971). Dopamine causes inhibition of prolactin production *in vitro* by both normal and tumorous pituitary cells, but inhibits growth hormone production only from tumorous material. Direct evidence for an effect of dopamine on prolactin secretion *in vivo* is given by the findings of Takahara *et al* (1976) that the infusion of dopamine into the portal circulation of animals results in suppression of prolactin release by the pituitary. Dopamine agonists, bromocriptine and apomorphine have a similar action to levodopa in causing suppression of plasma prolactin levels whilst chlorpromazine, tetrabenazine, pimozide and sulperide prevent the action of dopamine, and, given without levodopa, result in an elevation of plasma prolactin levels. Chlorpromazine acts as a *competitive* inhibitor of prolactin suppression by bromocriptine, giving further indirect evidence of an agonist role of dopamine in suppressing prolactin levels (Fluckiger, 1977).

Suppression of plasma prolactin levels in normal and hyperprolactinaemic subjects by levodopa or bromocriptine is rapid in onset, commencing within 15—30 minutes of oral dosage. In the case of levodopa, suppression lasts for 2—3 hours, and in the case of bromocriptine, for 6—8 hours following a single oral dose. A sustained reduction of elevated prolactin levels for 24 hours can be gained by giving bromocriptine 2.5—5 mg at 6 hour intervals. In both physiologic and non-physiologic hyperprolactinaemia, normal prolactin values may be achieved within a few hours of commencing therapy, but more usually this takes several days to realise. In hyperprolactinaemia (as in acromegaly), the hormone response to bromocriptine remains stable over 2—3 years of treatment and response fluctuations do not develop. During neither the initial nor the long term treatment of acromegaly and hyperprolactinaemia with dopamine stimulants is there any evidence that a change in receptor sensitivity occurs. Tolerance to bromocriptine does not develop in these disorders, and on the whole the same dosage remains effective throughout the period of treatment. Whether this is also true in parkinsonism, in which the pathology is progressive, is uncertain, although if anything levodopa dosage requires reduction rather than increase in the later years of treatment. This decrease may be necessary in some patients because of an increased prevalence of neuro-psychiatric side-effects.

In summary, there are important differences in the mechanisms of dopamine control of movement, growth hormone, and prolactin systems, although all three responses are likely to involve dopamine receptor stimulation. Possible differences in these systems have not yet been utilised in the treatment of acromegaly, hyperprolactinaemia and Parkinson's disease. However, if animal experiments are applicable to man, there may be sub-populations of dopamine receptors in movement systems, and it may prove possible to separate the actions of dopamine stimulants upon voluntary move ments and upon involuntary dyskinesias in parkinsonism.

Acknowledgements

We gratefully acknowledge the assistance of Mrs. P. Asselman and the most generous help of Sandoz towards the running of the Parkinson's Disease Clinic of King's College Hospital.

References

Cools, A.R. and Van Rossum, J.M. (1976). *Psychopharmacologia (Berlin)* **45**, 243.

Cotzias, G.C., Papavasiliou, P.S., Tolosa, E.S., Mendez, J.S. and Bell-Midura, M. (1976). *New England Journal of Medicine* **294**, 567.

Debono, A.G., Jenner, P., Marsden, C.D., Parkes, J.D., Tarsy, D. and Walters, J. (1977). *Journal of Neurology, Neurosurgery and Psychiatry* **40**, 162.

Delitala, G., Masala, A., Alagna, S. and Devilla, L. (1976). *IRCS Medical Science Library Compendium* **3**, 82.

Fluckiger, E. (1976). *In* "Pharmacological and Clinical Aspects of Bromocriptine" — proceedings of a symposium held at the Royal College of Physicians, London. MCS Consultants, Tunbridge Wells.

Johnson, A.M., Loew, D.M. and Vigouret, J.M. (1976). *British Journal of Pharmacology* **56**, 59.

Kolodny, H.D., Sherman, L., Singh, A., Kim, S. and Benjamin, F. (1971). *New England Journal of Medicine* **284**, 616.

Mars, H. and Genuth, S.M. (1973). *Clinical Pharmacology and Therapeutics* **14**, 390.

Marsden, C.D. and Parkes, J.D. (1976). *Lancet* i, 292.

Martin, J.B. (1973). *New England Journal of Medicine* **283**, 1384.

Muller, E.E., Saito, T., Arimura, A. and Schally, A.V. (1967). *Endocrinology* **80**, 109.

Pasteels, J.L., Danguy, A., Frerotte, M. and Ectors, F. (1971). *Annals of Endocrinology (Paris)* **32**, 188.

Rayfield, E.J., George, D.T., Eichner, H.L. and Hsu, T.H. (1975). *New England Journal of Medicine* **288**, 589.

Pycock, C.J. and Marsden, C.D. (1977). *Journal of Neurological Science* **31**, 113.

Schally, A.V., Arimura, A., Takahara, J., Redding, T.W. and Dupont, A. (1974). *Federation Proceedings* **33**, 237.

Shoulson, I., Glaubiger, G.A. and Chase, T.N. (1975). *Neurology (Minneapolis)* **25**, 1144.

Takahara, J., Arimura, A. and Schally, A.V. (1974). *Endocrinology* **95**, 462.

Thorner, M.O. (1975). *Lancet* i, 662.

Thorner, M.O. and Besser, G.M. (1976). *Postgraduate Medical Journal* **52**, Supplement 1, 71.

EXPERIMENTAL STUDIES OF DOPAMINE FUNCTION IN MOVEMENT DISORDERS

B. COSTALL and R.J. NAYLOR

Postgraduate School of Studies in Pharmacology, University of Bradford, Bradford, West Yorkshire, U.K.

Introduction

Observations in the clinic clearly indicate that drugs which are able to modify cerebral dopamine mechanisms are also able to modify motor function. Such drugs are therefore valuable tools for the experimental pharmacologist to investigate the role of dopaminergic mechanisms in motor control and to design new models for the detection of drugs with potential to relieve the symptoms of motor disorders. This is the approach taken in studies carried out in our laboratories and which we report in the present manuscript: these studies have aided our understanding of the site and mode of action of dopamine agonists and antagonists and have, at the same time, emphasised the limitation of such therapy. Last, but not least, data obtained from our own and other laboratories give some indication of future expectations both for the design of pharmacological tests and for the treatment of human disease states.

The Localisation of Cerebral Dopamine Systems

Almost all brain areas contain both noradrenaline and dopamine, but the relatively high concentrations of dopamine found in the olfactory tubercle, nucleus accumbens, caudate nucleus, globus pallidus, nucleus tractus diagonalis, nucleus septalis lateralis, lateral and central nucleus of the amygdala, median eminence and ventral tegmental area, and the modest amounts in the substantia nigra and cingulate, entorhinal and frontal cortex (Versteeg *et al*, 1976) would suggest a possible direct transmitter role within these areas. Histochemical visualisation of the fluorescence of biogenic amines induced under specific conditions initially indicated two distinct ascending telencephalic dopamine projections, the nigrostriatal and the mesolimbic systems (Ungerstedt, 1971). More recently, however, studies on the innervation to the cortex have indicated a further dopamine-containing 'mesocortical' system, and possible cross connections between the different systems (Fig. 1). However, the tuberoinfundibular dopamine systems and the dopamine mechanisms within the area postrema would appear to be anatomically discrete.

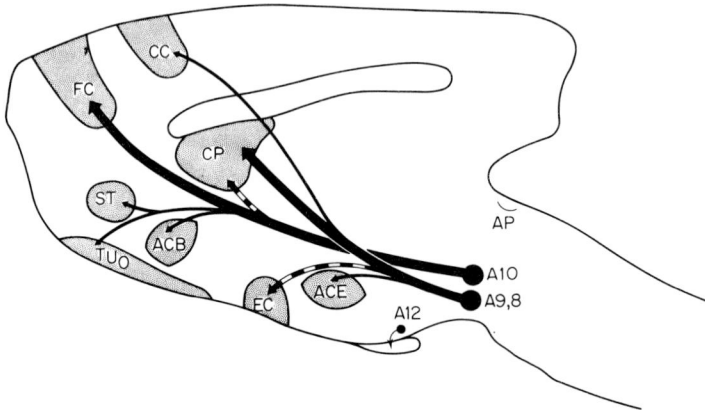

Fig. 1 Diagrammatic representation of the major dopamine systems in the rat brain. Solid lines indicate established neuronal pathways, broken lines indicate possible further pathways. ACB – nucleus accumbens, ACE – central nucleus of the amygdala, AP – area postrema, CC – cingulate cortex, CP – caudate-putamen, EC – entorhinal cortex, FC – frontal cortex, ST – stria terminalis, TUO – tuberculum olfactorium. The hypothalamic tuberoinfundibular system is indicated as A12. Areas are designated A8, A9, A10 and A12 according to the studies of Ungerstedt and colleagues. The diagram has been compiled by reference to Ungerstedt (1971), Berger *et al* (1976) and Simon *et al* (1976).

The Functional Role of Cerebral Dopamine Systems in Motor Control

At the present time there is little evidence to suggest that the dopamine mechanisms within the area postrema may modify motor function, but this lack of evidence relates to lack of experimentation, and some interesting observations have been made (Lichtensteiger and Lienhart, 1977): we must await further knowledge before judging the possible importance of the dopamine function within this brain area. In contrast, dopamine agonists and antagonists have a clear potential both to influence neuroendocrine function and to cause marked changes in motor performance. However, at this time the available data are considered to be too general in nature to allow the formulation of any hypothesis as to the possible role of hypothalamic systems in normal or abnormal motor activities. Therefore, the present chapter will be restricted to a consideration of the roles of striatal, mesolimbic and mesocortical systems in motor control.

The role of these systems in experimental animals has been investigated by applying the intracerebral injection and brain lesion

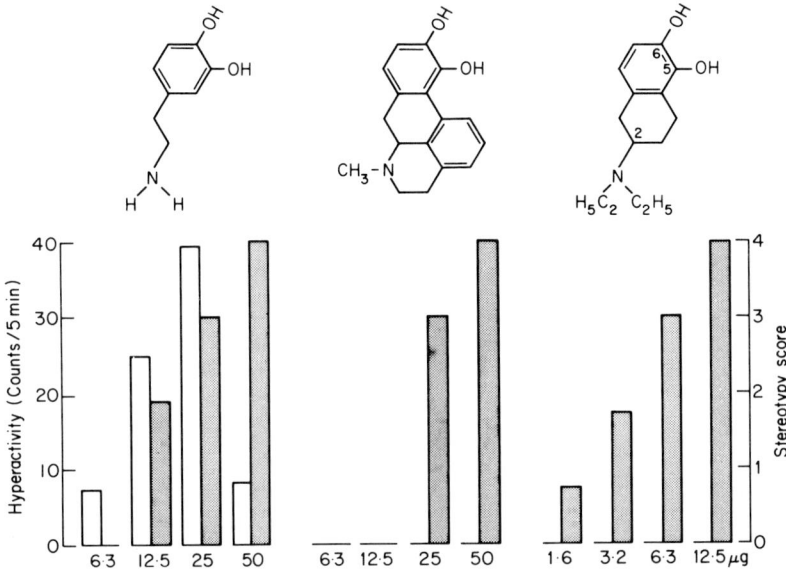

Fig. 2 Hyperactivity and stereotyped behaviour induced by dopamine, apomorphine and 2-(N, N-diethyl)amino-5, 6-dihydroxytetralin injected bilaterally into the caudate-putamen of rats pretreated (2 hr) with 100 mg/kg ip nialamide. See Costall *et al* (1977b) for details of the methodology. Briefly, drugs were administered bilaterally into the centre of the caudate-putamen complex in a volume of 1 or 2 ul (the unilateral dose is indicated in ug). The intensity of stereotypy was scored according to a simple system, 0 = no stereotypy, 1 = periodic sniffing and/or repetitive head and limb movements, 2 = continuous sniffing and/or repetitive head and limb movements, 3 = periodic gnawing, biting or licking, 4 = continuous gnawing, biting or licking. Hyperactivity was measured by placing the animals in individual perspex boxes equipped with photocell units: the number of light beam interruptions was recorded at 10 min intervals and counts/5 min are used to indicate the intensity of hyperactivity. The mean maximum response is shown for both stereotyped behaviour (hatched columns) and hyperactivity (open columns). 6−8 rats were used at each dose of drug. Standard errors on values presented are all less than 15% of the means.

techniques. Such studies have indicated that striatal dopamine system(s) play an important role in the modulation of 'stereotyped' motor responses. Many studies using the rat, cat or monkey have shown that intrastriatal dopamine injection can induce repetitive motor behaviour which is invariably characterised by peri-oral movements, biting, gnawing or chewing (Costall *et al*, 1974; Cools *et al*, 1975, 1976). In addition to these stereotyped responses, and dependent upon the precise experimental conditions, a more generalised hyperactivity may be induced (Costall and Naylor, 1976c) (Fig. 2). Initially these observations appear to contrast with the role of mesolimbic dopamine since, for example, many authors have shown that the primary response to dopamine injected into the mesolimbic nucleus accumbens is the production of hyperactivity, with stereotypy contributing a minor component to the overall effects on motor function (Pijnenburg and Van Rossum, 1973; Costall and Naylor, 1975a; Jackson *et al*, 1975). However, recent intra-cerebral injection studies have indicated that this may be too simplistic a view. We have used potent dopamine agonists from a series of N-alkylated 2-aminotetralins, phenylethylamines and N-propylnorapomorphine, and have clearly demonstrated that the nucleus accumbens can sustain a stereotyped biting (Figs. 3 and 4).

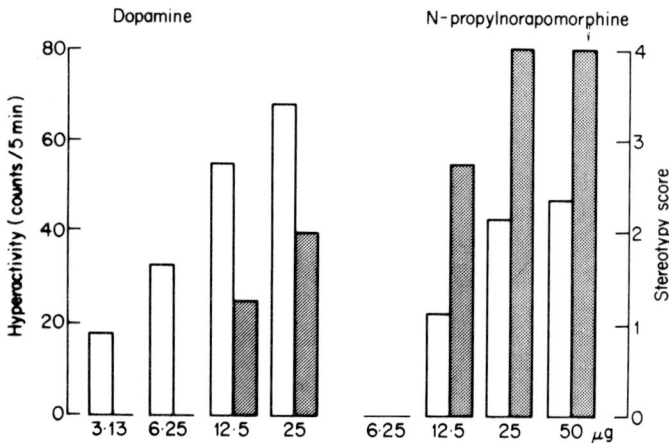

Fig. 3 Hyperactivity and stereotyped behaviour induced by dopamine and N-propylnor-apomorphine injected bilaterally into the nucleus accumbens of rats pretreated (2 hr) with 100 mg/kg ip nialamide. Drugs were administered into the centre of the nucleus accumbens in a volume of 1 ul (the unilateral dose is indicated in ug). The mean maximum response is shown for both stereotyped behaviour (hatched columns) and hyperactivity (open columns). See legend to Fig. 2 for more details of the methodology. 6−8 rats were used at each dose of drug. Standard errors on values presented are all less than 14% of the means.

Furthermore, certain of these dopamine agonists, for example, 2-(N, N-dipropyl) amino-5,6-dihydroxytetralin, induced biting *without* causing hyperactivity. The implication of these observations is far-reaching since it necessitates a dissociation between the dopamine mechanisms mediating stereotypy and hyperactivity: the terminology of 'dopamine mechanisms' may therefore require careful qualification. This concept is expanded even further by the observation that derivatives of 2-aminotetralin and apomorphine may differentially activate those mechanisms in the extrapyramidal and mesolimbic systems which modulate stereotyped motor behaviour; stereotypy may be induced by certain drugs when injected into the nucleus accumbens but not when similarly applied to the striatum (Costall

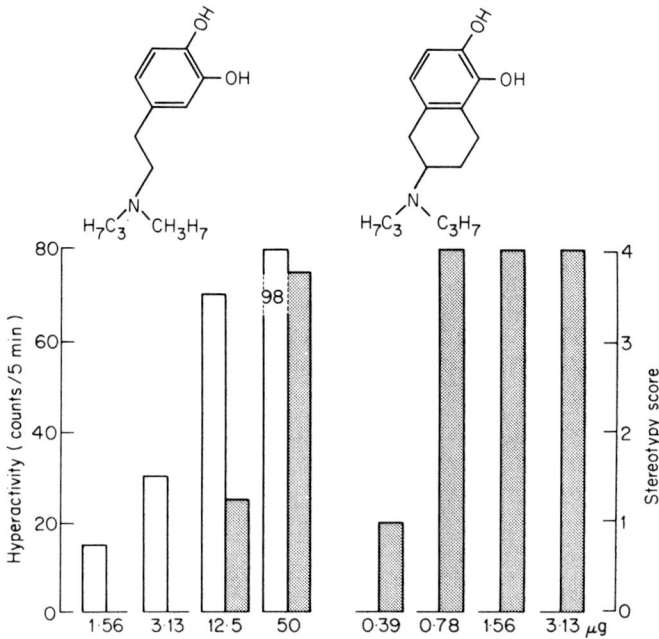

Fig. 4 Hyperactivity and stereotyped behaviour induced by N,N-dipropyldopamine and 2-(N,N-dipropyl)amino-5,6-dihydroxytetralin injected bilaterally into the nucleus accumbens of rats pretreated (2 hr) with 100 mg/kg ip nialamide. Drugs were administered into the centre of the nucleus accumbens in a volume of 1 ul (the unilateral dose is indicated in ug). The mean maximum response is shown for both stereotyped behaviour (hatched columns) and hyperactivity (open columns). See legend to Fig. 2 for more details of methodology. 6−8 rats were used at each dose of drug. Standard errors on values presented are all less than 12% of the means.

et al, 1977a, b). However, it is implicit in the formation of hypotheses of differing dopamine mechanisms that the observations using purported dopamine agonists should be supported by the use of selective antagonists. In preliminary experiments we have shown that some neuroleptics can differentially reduce dopamine hyperactivity induced from mesolimbic or striatal areas (Costall and Naylor, 1976c). In these 'hyperactivity models' (induced by the injection of dopamine into the striatum or nucleus accumbens) sulpiride and clozapine, for example, are relatively more effective antagonists of the dopamine response from the mesolimbic area whilst metoclopramide and other agents are more effective in reducing the response from the striatum (Figs. 5 and 6). Based on these observations one would tentatively extrapolate to stereotyped motor behaviour and postulate that this may also be differentially antagonised. However, we have not as yet

Nucleus Accumbens   Caudate-Putamen

Sulpiride   Sulpiride

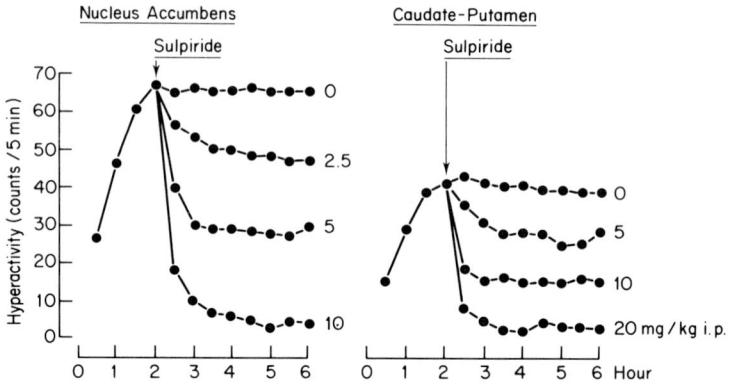

Fig. 5 Antagonism by sulpiride of the hyperactivity induced by 50 ug (1 ul) dopamine injected into the nucleus accumbens or 25 ug (2 ul) dopamine injected into the caudate-putamen. Rats were pretreated (2 hr) with 100 mg/kg ip nialamide. The arrows indicate the times of injection of sulpiride, and the doses administered are indicated in mg/kg ip. Hyperactivity was measured as indicated on the legend to Fig. 2. See Costall and Naylor (1976c) for more detailed methodology. Values are the means of responses of 6−10 rats. Standard errors are all less than 17% of the means.

performed the relevant experiments. Nevertheless, we have attempted to dissociate the stereotypy mechanism within the striatum from the hyperactivity mechanism within the striatum.

At this stage of the studies some complexities in terminology became apparent and it was found necessary to expand on the

Nucleus Accumbens   Caudate-Putamen

Metoc.   Metoc.

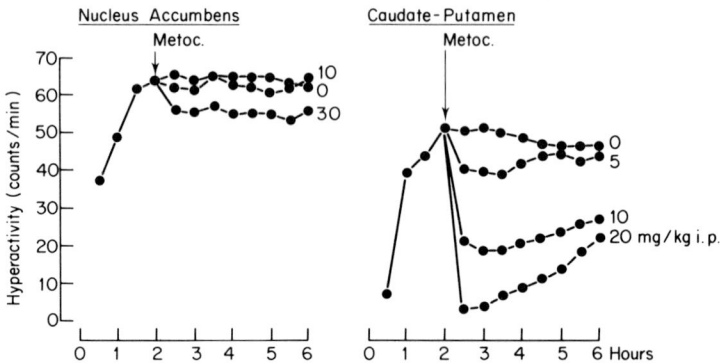

Fig. 6 Effect of metoclopramide on the hyperactivity induced by 50 ug (1 ul) dopamine injected into the nucleus accumbens or 25 ug (2 ul) dopamine injected into the caudate-putamen. Rats were pretreated (2 hr) with 100 mg/kg ip nialamide. The arrows indicate the times of injection of metoclopramide, and the doses administered are indicated in mg/kg ip. Hyperactivity was measured as indicated on the legend to Fig. 2. See Costall and Naylor (1976c) for more detailed methodology. Values are the means of responses of 6−10 rats. Standard errors are all less than 14% of the means.

terminology so commonly applied to oral movements (stereotypy). In the clinic, a distinction can be made between stereotyped and dyskinetic movements (Marsden *et al*, 1975), and we would forward the concept that perioral movements induced in animals may be stereotypic or dyskinetic in nature (Costall and Naylor, 1976a). This concept was formulated on the basis of observations using dopamine agonists and antagonists of differing pharmacological activity spectra, initially applied to the guinea-pig. We developed the 'guinea-pig model' for the detection of anti-dyskinetic agents. In this species intrastriatal dopamine induces a pronounced hyperactivity concomitant with peri-oral effects (Costall *et al*, 1975). All classical neuroleptic agents appear to reduce the hyperactivity but the peri-oral movements are generally resistant, except to the actions of three agents, oxiperomide, pimozide and tiapride (Costall and Naylor, 1975b) (Figs. 7 and 8). Oxiperomide is the most potent antidyskinetic agent so far evaluated but has a clear ability to reduce overall motor performance. Pimozide, whilst less

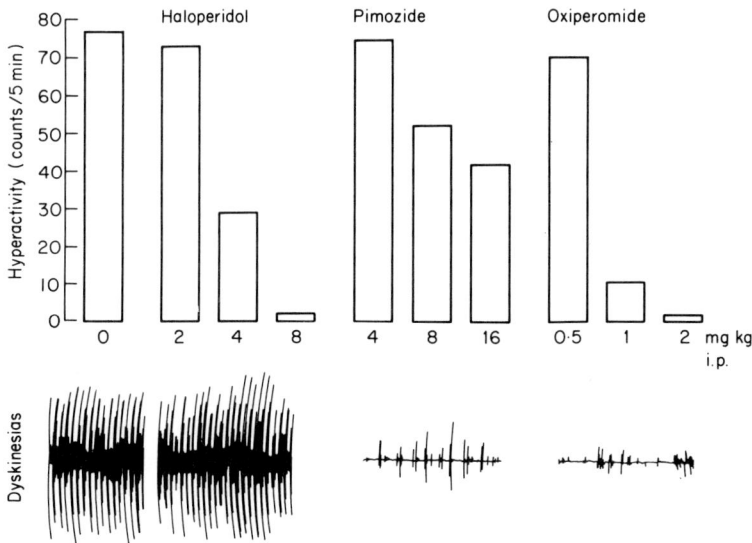

Fig. 7 Effect of haloperidol, pimozide and oxiperomide on the hyperactivity and dyskinesias induced in the guinea-pig by bilateral intrastriatal injections of 100 ug (2 ul) dopamine. Animals were pretreated (2 hr) with 75 mg/kg ip nialamide. The potential antagonists were administered when dyskinesias were established approximately 2 hr after the dopamine injections. Hyperactivity was measured in the same way as described for the rat (legend, Fig. 2) and the means of the maximum effects observed in 6–10 animals are shown for each dose of drug (doses shown in mg/kg ip). Standard errors on these values were less than 18% of the means. Abnormal dyskinetic movements were observed visually and were also recorded as oscillations on a Grass polygraph. Typical recordings are shown for each drug. For more detailed explanations of the methodology used see Costall and Naylor (1975b).

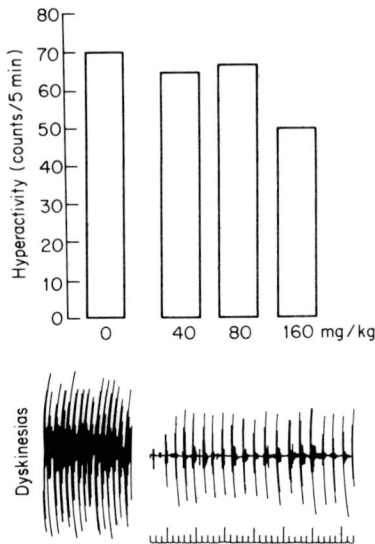

Fig. 8 Effect of tiapride on the hyperactivity and dyskinesias induced in the guinea-pig by bilateral intrastriatal injections of 100 ug (2 ul) dopamine. Animals were pretreated (2 hr) with 75 mg/kg ip nialamide. Tiapride was administered when dyskinesias were established, approximately 2 hr after the dopamine injections. Hyperactivity was measured as described in the legend to Fig. 2 for the rat and the means of the maximum effects observed in 6–8 animals are shown for each dose of tiapride (doses shown in mg/kg ip). Standard errors on these values range between 12 and 19% of the means. No significant (P > 0.05) antagonism was observed, even at 160 mg/kg ip. Abnormal dyskinetic movements were observed visually and were also recorded as oscillations on a Grass polygraph. Typical recordings are shown. Generally, the low amplitude recordings represent more restricted abnormal movements such as peri-oral dyskinesias, and the high amplitude recordings correlated with abnormal movements of the body, for example, jerking of the entire body or sudden movements of the limbs (myoclonus). For tiapride, the abnormal facial movements were preferentially abolished (compare with the effects of, for example, oxiperomide, Fig. 7). For more detailed explanations of the methodology used see Costall and Naylor (1975b).

potent, can also effectively abolish dyskinesias with a somewhat reduced effect on locomotor activity. Tiapride has an ability to antagonise peri-oral dyskinesias without inhibiting general locomotor activity. These observations again indicate that the components of the dopamine response from a specific brain region may be differentially antagonised and may therefore involve the activation of different dopamine mechanisms. These, and other observations, have ultimately led to the postulation of two different types of dopamine receptor. Further, to adequately explain our behavioural observations this concept has been expanded to indicate not only the possibility of differences between the dopamine mechanisms within an area, but also a difference in the dopamine mechanisms which are located in different brain areas and are capable of mediating the same motor response (Costall *et al*, 1977b).

The above hypotheses may be applied to extrapyramidal and mesolimbic brain areas, but as yet the functions of the mesocortical system have not been the subject of any extensive studies, and it would be far too speculative to attempt an extrapolation to this brain region. Data on the functional role of the mesocortical system have been obtained somewhat indirectly by lesions of the area A10 (Galey *et al*, 1977) which largely innervates the forebrain and cortical limbic structures. It has been suggested that activation of this system may exert a general inhibitory effect on behaviour. Certainly, this is a brain region which must be subjected to more extensive investigation if we are to understand more fully the nature of the actions of dopamine agonists and antagonists on motor function, and hence further elucidate the cerebral dysfunctions in motor disorders.

Clinical Implications of Experimental Studies on Dopamine Function in Motor Control

In considering the clinical implications of the experimental studies we firstly assume that the dopamine mechanisms which mediate the general locomotor activity following intracerebral injection into mesolimbic or extrapyramidal areas in animals have an analogous substrate(s) in man which mediates an anti-akinetic action. Secondly, we assume that the stereotyped peri-oral movements observed after the injection of dopamine and dopamine agonists into dopamine-containing areas of the animal brain may relate to an excessive stimulation of normal or latent dopamine mechanisms which have an analogous substrate(s) involved with the mediation of abnormal peri-oral/oro-bucco-lingual motor behaviour in the clinic. It is not an essential requirement that the dopamine mechanisms in animals and man should be the same, nor is it prerequisite that the mechanisms are located in precisely the same anatomical substrate (indeed, this would not even appear possible when one considers the differences in cerebral anatomy of the rat and man; taking the nucleus accumbens as a pertinent example, this area is anatomically distinct from the striatum in the rat, but probably forms a component of the rostral caudate complex in man. The essential points are that the dopamine receptors and/or mechanisms which mediate locomotor activity or peri-oral dyskinesias firstly, are discrete and, secondly, may possess different pharmacological characteristics. The concept of two different dopamine receptor mechanisms, which have been termed DA-1 (for hyperactivity induction) and DA-2 (for dyskinesia induction) (Costall and Naylor, 1975b; Cools *et al*, 1975), considerably expands the traditional concept in which it is tacitly assumed that a single type of dopamine receptor mediates the hyperactivity response on modest stimulation and the dyskinetic manifestations on marked stimulation.

It is appreciated that many neurotransmitter mechanisms are relevant to a control of normal and abnormal motor function, but the ability of 1-dopa and bromocriptine to induce dyskinesias (see review by Marsden *et al*, 1975; Parkes *et al*, 1976), and the ability of other dopamine agonists to modify existing dyskinetic phenomena (Gessa *et al*, 1972; Tolosa and Sparber, 1974), strongly supports an important involvement of dopamine mechanisms. Dopamine agonists are widely used to alleviate akinesia, but the appearance of dyskinesias constitutes one of the most limiting aspects of dopamine agonist therapy. Indeed, dyskinesias (DA-2 stimulation) can clearly develop in the absence of an anti-akinetic effect. Such an observation may be expected if the experimentally derived concept of different dopamine mechanisms mediating the two effects can be extrapolated to man. It would follow that it may also be possible to design agents to selectively stimulate the DA-1 mechanisms and thus derive a selective anti-akinetic drug, but at the present time this remains a theoretical possibility. Conversely, it may also be possible to design other agents with a partial agonist/antagonist activity to block the DA-2 system and have a potential antidyskinetic activity. It may be that such a potential partly accounts for the clinical observations that some patients with dyskinesias experience some relief from their abnormal involuntary movements when treated with apomorphine, N-propylnorapomorphine or lergotrile (Duby *et al*, 1972; Cotzias *et al*, 1976; Lieberman *et al*, 1976).

An associated and challenging problem is to dissociate the ability of dopamine agonists to modify motor dysfunction from an ability to excessively stimulate those dopamine systems, probably present within mesolimbic and/or mesocortical circuits, which modulate thought and emotional behaviour. An improper stimulation of these mechanisms, as evidenced by the indiscriminate cerebral action of 1-dopa, results in psychomotor agitation or schizophreniform symptomatology. Bromocriptine and piribedil may have a similar action (Chase *et al*, 1974; Parkes *et al*, 1976) but it is interesting that the classical dopamine agonist, apomorphine, fails to induce psychotic disturbance or psychomotor agitation in man (Angrist *et al*, 1975);

indeed, a recent report suggests that apomorphine may actually *reduce* psychotic disturbance in man (Rotrosen *et al*, 1976). This may correlate well with the animal observations: apomorphine will actually *reduce* the hyperactivity induced by dopamine from the nucleus accumbens of rat (Costall and Naylor, 1976b). It is suggested, therefore, that a dissociation between a stimulation of striatal and higher dopaminergic centres may be a realistic concept, and worthy of further study.

The general concepts so far discussed for the activation of different dopamine mechanisms also apply to their antagonism. The classical dopamine antagonists include, of course, the neuroleptic agents, which still remain as the standard therapy for a number of abnormal involuntary movement disorders. However, one great disadvantage of such therapy is that the neuroleptics concurrently reduce normal movements, frequently to an unacceptable degree. This is clearly to be expected in so far as the classical neuroleptics, the phenothiazines, butyrophenones and other series, exert an apparently non-selective inhibition on all cerebral dopamine systems. Thus, inhibition of the DA-1 hyperactivity mechanisms presumably results in the neuroleptic akinesia and, with increasing dosage, a full pseudo-parkinson syndrome may develop as the result of additional effects, direct or indirect, on other neurotransmitter mechanisms. Since very large doses of neuroleptic agents are required to inhibit the DA-2 dyskinetic mechanism, such antidyskinetic therapy is invariably associated with a poverty of movement. Clearly, it would be advantageous to develop neuroleptic agents with a selectivity of action on the DA-1 and DA-2 mechanisms. A selective DA-1 antagonist may, theoretically, be of particular value in reducing excessive psychomotor agitation or mania, whereas a selective DA-2 antagonist may have a specific antidyskinetic potential.

Our studies in the guinea-pig have indicated three potential DA-2 antagonists, pimozide, oxiperomide and tiapride. However, a complete specificity for the DA-2 mechanisms cannot be claimed for all these agents. Pimozide and oxiperomide can undoubtedly affect DA-1 mechanisms and reduce locomotor activity, but the dosage differential for DA-1 and DA-2 antagonism is narrow for these agents, and markedly contrasts with that of the classical neuroleptics whose affinity appears to be most preferentially DA-1. In contrast, even at high pharmacological doses, we have not as yet obtained any data to indicate an action of tiapride on DA-1 mechanisms. On the basis of this animal experimentation we would suggest that these three agents should exert more selective antidyskinetic activity in the clinic than at present shown by the classical neuroleptics, and evidence is now available to confirm our hypotheses. Pimozide has been shown to abolish drug induced dyskinesias (1-dopa and neuroleptic induced) without necessarily inducing an unacceptable degree of motor impairment (Claveria *et al*, 1975; Tarsy *et al*, 1975), although it cannot be ignored that some patients do appear to respond more adversely. Preliminary clinical studies have also established the antidyskinetic potential of oxiperomide (1-dopa induced dyskinesias), although careful attention appears to be necessary to avoid interference with normal motor behaviour, as one would predict from the animal studies (Marsden, personal communication). Tiapride has also been shown to possess antidyskinetic properties in the clinic, with a spectrum of action which could be predicted on the basis of animal observations (L'Hermitte, personal communication).

Conclusions

It is suggested that there is now sufficient evidence available from the use of dopamine agonists and antagonists in animal models to put forward differential roles for the major cerebral dopamine systems, and even different dopamine receptor mechanisms within those systems, for the modulation of various aspects of motor behaviour. Preliminary clinical observations encourage an extrapolation of these findings to man, and the hypothesis has been put forward that the cerebral dopamine mechanisms may differentially malfunction in movement disorders. Initial studies have offered some confirmation of the possibility that the different dopamine mechanisms may be stimulated or antagonised by drugs with relative specificity to alleviate the symptoms of motor dysfunction, and it is considered that the future should bring the development of new drugs having specific actions on the different dopamine mechanisms and, therefore, a specificity of effect for the treatment of movement disorders for which, so far, a satisfactory drug treatment has not been available.

Acknowledgements

This work was supported by the Medical Research Council.

References

Angrist, B., Thompson, H., Shopsin, B. and Gershon, S. (1975). *Psychopharmacologia (Berl.)* **44**, 273–280.

Berger, B., Thierry, A.M., Tassin, J.P. and Moyne, M.A. (1976). *Brain Res.* **106**, 133–145.

Chase, T.N., Woods, A.C. and Glaubiger, G.A. (1974). *Arch. Neurol.* **30**, 383–386.

Claveria, L.E., Teychenne, P.F., Calne, D.B., Haskayne, L., Petrie, A., Pallis, C.A. and Lodge-Patch, I.C. (1975). *J. Neurol Sci.* **24**, 393–401.

Cools, A.R., Hendriks, G. and Korten, J. (1975). *J. Neural. Transmission* **36**, 91–105.

Cools, A.R., Struyker-Boudier, H.A.J. and Van Rossum, J.M. (1976). *Eur. J. Pharmac.* **37**, 283–293.

Corsini, G.U., Del Zompo, M., Cianchetti, C., Mangoni, A. and Gessa, G.L. (1976). *Psychopharmacology* **47**, 169–173.

Costall, B. and Naylor, R.J. (1975a). *Eur. J. Pharmac.* **32**, 87–92.

Costall, B. and Naylor, R.J. (1975b). *Eur. J. Pharmac.* **33**, 301–312.

Costall, B. and Naylor, R.J. (1976a). *Eur. J. Pharmac.* **36**, 423–429.

Costall, B. and Naylor, R.J. (1976b). *J. Pharm. Pharmac.* **28**, 592–595.

Costall, B. and Naylor, R.J. (1976c). *Eur. J. Pharmac.* **40**, 9–19.

Costall, B., Naylor, R.J. and Pinder, R.M. (1974). *J. Pharm. Pharmac.* **26**, 753–762.

Costall, B., Naylor, R.J. and Pinder, R.M. (1975). *Eur. J. Pharmac.* **31**, 94–109.

Costall, B., Naylor, R.J., Cannon, J.G. and Lee, T. (1977a). *Eur. J. Pharmac.* **41**, 307–319.

Costall, B., Naylor, R.J., Cannon, J.G. and Lee, T. (1977b). *J. Pharm. Pharmac.* **29**, 337–342.

Cotzias, G.C., Papavasiliou, P.S., Tolosa, E.S., Mendez, J.S. and Bell-Midura, M. (1976). *New Eng. J. Med.* **294**, 567–572.

Duby, S.E., Cotzias, G.C., Papavasiliou, P.S. and Lawrence, W.H. (1972). *Arch. Neurol. (Chic.)* **27**, 474–480.

Galey, D., Simon, H. and Le Moal, M. (1977). *Brain Res.* **124**, 83–97.

Gessa, R., Tagliamonte, A. and Gessa, G.L. (1972). *Rivista di Farmacologia e Terapia* **111**, 423–427.

Jackson, D.M., Anden, N.-E. and Dahlstrom, A. (1975). *Psychopharmacologia (Berl.)* **45**, 139–149.

Lichtensteiger, W. and Lienhart, R. (1977). *Nature* **266**, 635–637.

Lieberman, A.N., Kupersmith, M., Estey, E. and Goldstein, M. (1976). *The Lancet*, 515–516.

Marsden, C.D. Tarsy, D. and Baldessarini, R.J. (1975). *In* "Seminars in Psychiatry" p. 219. Raven Press, New York.

Parkes, J.D., Debono, A.G. and Marsden, C.D. (1976). *Neurol. Neurosurg. and Psychiat.* **39**, 1101–1108.

Pijnenburg, A.J.J. and Van Rossum, J.M. (1973). *J. Pharm. Pharmac.* **25**, 1003–1005.

Rotrosen, J., Angrist, M.B., Gershon, S., Sachar, E.J. and Halpern, F.S. (1976). *Psychopharmacology (Berl.)* **51**, 1–7.

Simon, H., Le Moal, M., Galey, D. and Cardo, B. (1976). *Brain Res.* **115**, 215–231.

Tarsy, D., Parkes, J.D. and Marsden, C.D. (1975). *J. Neurol. Neurosurg. and Psychiat.* **38**, 331–335.

Tolosa, E.S. and Sparber, S.D. (1974). *Life Sci.* **15**, 1371–1380.

Ungerstedt, U. (1971). *Acta. Physiol. Scand. Suppl.* **367**, 1–48.

Veersteeg, D.H.G., Van Der Guten, J., De Jong, W. and Palkovits, M. (1976). *Brain Res.* **113**, 563–574.

HUNTINGTON'S CHOREA

EDWARD D. BIRD

*Neurochemical Pharmacology Unit, Department of Pharmacology,
The Medical School, Cambridge, U.K.*

Incidence

Every physician recalls, from his medical school training, the story of immigrants from East Anglia who landed in Salem, Massachusetts in 1632, and who were traced as ancestors to a very large number of choreic families in the United States (Vessie, 1932). Although there is some controversy over the actual family names used in that study (Hans and Gilmore, 1969; Caro, 1977b) we know that this inherited disorder must have been carried to New England during the massive migration from the British Isles in the 15th century, and therefore the story keeps physicians aware of this autosomal dominant disorder that is inherited by 50% of the offspring.

There may be a greater incidence of Huntington's chorea in East Anglia than the rest of the U.K., since Caro (1977a) has found 9.24 cases per 100,000 population in this region. Heathfield and McKenzie (1971) found 7 cases per 100,000 in Bedfordshire, which was an increase over the 5.1 per 100,000 found by Pleydell (1954) in Northamptonshire. This steady increase in reported incidence with time is I suspect due to an increased awareness of the disorder, and future assessments in other regions of the U.K. and other parts of the Western World will probably turn out to be the same as that found in East Anglia.

However, the problems that the disorder creates for the family physician are greater than these figures might suggest, for they are compounded by the late onset of the chorea in middle life after the reproductive period. By the time chorea appears there will be a large number of family members who with their spouses and 'at risk' children will be anxious and need counselling. In addition, the spouses of those already afflicted become physically exhausted from the 24 hour nursing care required, and will need a physician's attention. A realistic incidence for the number of family members involved in one way or another with this disorder thus reaches almost 1 per 1,000, and the problem is therefore likely to present itself in every medical practice.

Clinical Features and Genetics

The disorder may present at a very early age with convulsions. The earliest case that I have been associated with died at the age of 6 years. The child had been admitted to a children's home after having had uncontrolled convulsions for a number of years. Later his father developed choreiform movements, and the diagnosis of juvenile chorea was considered. The post-mortem examination of the brain of this child was typical for Huntington's chorea. The juvenile choreic may present with rigidity, making the diagnosis difficult especially if both parents are free of movements. In juvenile cases the abnormal gene will have been inherited from the father four times more frequently than from the mother (Merritt *et al*, 1969).

At adolescence and in the third decade 'at risk' cases may present with behavioural abnormalities. A number have appeared psychotic and have been admitted to psychiatric hospitals with the diagnosis of schizophrenia, and only about 10 years later when the choreiform movements appear has the diagnosis been changed. Psychotic features are more prominent in those families that have an earlier onset of choreiform movement.

The most common and typical presentation for Huntington's chorea is slight twitching in the face or hands, starting between the ages of 35 and 50 years. The abnormal movements progress over some 15 years, with the need for 24 hour nursing care during the last few years in bed; when death finally occurs it is usually secondary to bronchopneumonia. Dementia is commonly present, and becomes more pronounced towards the latter stages of the disease. Once the dementia develops the downhill progress of the disease is more rapid.

When the onset of choreiform movements is in the 6th, 7th or 8th decade the progress of the disease is very slow, and it is rare for such patients to have dementia. Many of these patients have their chorea for more than 20 years, and usually die from natural causes.

I would propose the following explanations for the various clinical manifestations of the disease. There are probably sex-related modifying factors that counterbalance the genetic 'biochemical defect' characteristic for Huntington's disease. The juvenile choreic, with very little positive modifying factor, has a rapid rate of cell death: the disease presents in early life and rapidly progresses to death in less than 15 years. The gross pathological examination reveals an extreme degree of basal ganglia atrophy with often only a thin ribbon of caudate tissue left. The rigidity commonly associated with juvenile chorea is probably due to the extreme loss of cells in the basal ganglia, including those post-synaptic cells that have receptors for dopamine. We have found no decrease in dopamine concentration in the post-mortem striatum, and in fact in those cases that have rigidity prior to death the dopamine concentration is greater than normal (Bird and Iversen, 1974a). Therefore, the cause of the rigidity is different from that of Parkinson's disease, where the dopamine concentration in the striatum is decreased. Cortical cell loss also occurs and may be the cause of convulsions at this age.

The inheritance of a greater amount of positive modifying factor would delay the rate of cell death such that it might take more than 60 years for enough cells to degenerate in the basal ganglia and produce chorea. The rate of disease progress would be slow and the cortical cell loss, which is hardly ever as severe as the basal ganglia loss, would not be great enough to produce dementia.

The age of onset for chorea is generally the same throughout several family generations except when the disease is passed on by an affected male. As mentioned previously, occasionally the offspring of affected males may present with the disease in childhood. We have also noted that when the offspring of affected males are compared with the offspring of affected females, the age of onset and age of death will be earlier than in an affected father but at the same age as an affected mother (Bird *et al*, 1974). This suggests that the affected male may lack some positive genetic modifying factor that is provided by the affected female.

Pathology

Over the last six years we have collected over 200 choreic post-mortem brains. At autopsy one half of each brain is placed in formalin for histopathological examination by Professor Corsellis and his colleagues. The second half of the brain is frozen for future dissection and neurochemical assay.

Almost all of the brains received have shown atrophy of the caudate nucleus, putamen and globus pallidus, which is typical for this disorder. The basal ganglia lose more than 50% of their original weight. The whole brain loses between 15 and 25% of its weight, and this is largely due to cortical cell loss. The sulci are wide and the ventricles appear large due to the cortical cell degeneration, which occurs mainly in layers 3, 5 and 6. The amygdala and hippocampus, however, are often minimally affected by the atrophic changes. The brainstem and the cerebellum become atrophic but this tends to be more variable (Corsellis, personal communication).

The frozen hemisphere is dissected while frozen into a large number of neuroanatomical regions and stored in a deep freeze until analysed for neurochemical substances.

Neurochemistry

In addition to carrying out our own investigations, we have been able to share tissue with other investigators, and this has provided additional neurochemical information on the disorder. Perry *et al* (1973) originally showed that there was a decrease in gamma aminobutyric acid (GABA) in the basal ganglia of post-mortem choreic brain. The biosynthetic enzyme for GABA, glutamic acid decarboxylase (GAD) was also markedly decreased in this region of the brain but not in the cerebral frontal cortex (Bird *et al*, 1973) (see Table I). The mean activity of choline acetyltransferase (ChAc) in the basal ganglia of choreic brain was noted to be decreased by 50% (Bird *et al*, 1973). With a larger series of brains it was apparent that ChAc activities were normal in the basal ganglia of some choreic brains and markedly decreased in others (Bird and Iversen, 1974b). It now appears that the decrease in ChAc must occur in the late stages of this disease, because normal ChAc activity is found in the brains of those cases that die from natural causes, before the terminal stage of their disease. Tyrosine hydroxylase (T-OH), the biosynthetic enzyme for dopamine, was increased in the substantia nigra (see Table I).

The inter-relationship between the gabanergic and dopaminergic neurons can be best visualised by the schematic drawing in Fig. 1. The substantia nigra is divided into two regions, the more dorsal region being the zona compacta which is normally darkly pigmented and contains the cell bodies of the dopamine neurons whose axons form a pathway to the striatum. Dendrites extend from the dopamine cell body throughout both regions of the substantia nigra. The less pigmented ventral region of the substantia nigra, the zona reticulata, receives axons from the gabanergic cells in the striatum, and the terminals of these axons are in contact with the dendrites of the dopamine cells. The dopamine concentration in the zona compacta is normally twice that of the zona reticulata (Hornykiewicz, 1963) and the GABA concentration is two-fold higher in the reticulata zone than the compacta zone (Kanazawa and Toyokura, 1975).

In Huntington's chorea there is a decrease in GAD activity due to

TABLE I

*Glutamic Acid Decarboxylase, Choline Acetyltransferase and Tyrosine Hydroxylase
in Control and Choreic Post-mortem Brain*

	Control	Huntington's Chorea	P
GAD[a] Caudate	43.1 ± 2.8 (111)	16.5 ± 2.0 (91)	< 0.001
Substantia Nigra	68.7 ± 9.13 (40)	33.38 ± 5.14 (33)	< 0.005
Frontal Cortex	38.9 ± 4.97 (15)	39.4 ± 4.33 (20)	N.S.
ChAc[b] Caudate	184.0 ± 9.5 (89)	105.0 ± 16.2 (60)	< 0.001
Frontal Cortex	1.9 ± 1.39 (22)	1.7 ± 1.31 (18)	N.S.
T-OH[c] Caudate	24.5 ± 4.49 (30)	34.4 ± 10.1 (23)	N.S.
Substantia Nigra	28.3 ± 3.57 (39)	75.21 ± 17.2 (40)	< 0.02

a = Glutamic Acid Decarboxylase (umol/h/g protein)
b = Choline Acetyltransferase (umol/h/g protein)
c = Tyrosine Hydroxylase (umol/h/g protein)
All values are means ± S.E.M. for the number of brains shown in parentheses.

the degeneration of the gabanergic cells in the striatum and thus of
their terminals, which are mainly in the zona reticulata close to the
dendrites of the dopamine neuron. The dendrites are thought to
have special receptors for GABA, and the loss of this neuro-inhibitory
transmitter in choreic tissue would increase dopaminergic activity.
Such increased activity would be reflected by abnormal movements,
just as the administration of excessive L-dopa to patients with
Parkinson's disease may produce abnormal movements.

The increased T-OH activity and dopamine concentration in the
substantia nigra may be a reflection of this loss of inhibitory control.
However, since the atrophy in the substantia nigra is due to a loss
of gabanergic rather than dopaminergic neurons the increased
dopaminergic activity in choreic tissue may be due to a relative
increase in the concentration of these remaining neurons.

Using radioactive labelled GABA, the ability of brain cell
membrances to bind GABA can be used to characterise the GABA
receptor. In the substantia nigra from choreic brain the GABA

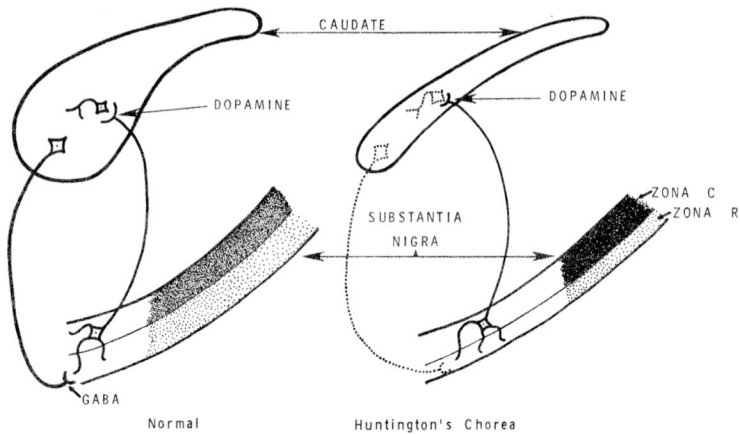

Fig. 1 Schematic representation of the striatal-nigral connections showing dopaminergic pathways from the zona compacta to the striatum and gabanergic pathways from the striatum to the zona reticulata in both normal and choreic brain.

receptor activity as measured by this binding technique is increased two-fold, suggesting that the GABA receptor is part of the dopamine cell (Enna *et al*, 1976).

The increase in dopaminergic activity in brain may also be reflected by alterations in hypothalamic activity; for example, dopamine stimulates the release of gonadotropic releasing factor (GRF) from the hypothalamus. The post-mortem hypothalamus from female choreics has been found to have an increased concentration of GRF (Bird *et al*, 1976). Female choreics have been noted to be more fertile than their non-choreic female siblings (Reed and Palm, 1951). Sexual precocity is fairly common, and males will often request that some medication be given to their choreic sexually overactive wives. Metrorrhagia also sometimes is seen and may require hysterectomy.

Plasma growth hormone (Podolsky and Leopold, 1974) and free fatty acids (Phillipson and Bird, 1977) are increased in patients with Huntington's chorea, findings that may be secondary to increased dopaminergic activity inhibiting somatostatin in the hypothalamus. Prolactin release is inhibited by dopamine, and there is some evidence for decreased plasma prolactin in patients with chorea (Hayden *et al*, 1977).

Pharmacology

In the hope that some clues to the biochemical defect in the disorder may be revealed by the response to pharmacologic agents, a large number of drug trials have been carried out. Many preliminary reports on the use of various drugs have indicated some decrease in choreiform movements, but follow up reports with the results of double blind trials have not appeared. The placebo effect is so great in this disease that the physician should be wary of drug successes unless an agent has been assessed by the double blind method.

Drugs that are given to increase the acetylcholine concentration in the brain such as dimethylaminoethanol (Deanol), choline or lecithin probably do not hold much promise since the cholinergic cells appear to die fairly late in the disease.

Agents given to increase GABA concentrations are being tried. Sodium valproate (Epilim) blocks enzymes that metabolise GABA, but preliminary trials with choreic patients are not encouraging. If a suitable agent can be found to increase brain GABA it will probably cause generalised neuroinhibition through the brain and not selectively in the basal ganglia. Choreic patients like to keep as alert as possible, especially when walking.

Until we can find some way to prevent the gabanergic cell dying I think we will be unlikely to alter the course of this disease. Drugs which interfere with the re-uptake of dopamine such as tetrabenazine (Nitoman) have been found helpful in reducing violent movements. As might be expected, patients may become depressed with tetrabenazine. This drug should probably be reserved for use when the movements are severe, and its dose should then be titrated, starting with a half tablet (12½ mg) three or four times a day to avoid depression. For the occasional violent patient the phenothiazine or butyrophenone drugs are helpful.

Interest in Huntington's chorea is no longer confined to East Anglia and the New England States but is world wide, now that the disorder is more frequently being recognised. The results of the neurochemical studies being carried out on choreic brain tissue will have an application to a better understanding of this disorder as well as to a number of other neurological and psychiatric disorders. This clinical and scientific interest provides some hope and encouragement to the many young members of our community that have already inherited the abnormal gene of this tragic disease.

References

Bird, E.D., Caro, A.J. and Pilling, J.B. (1974). *Ann. Hum. Genet. (Lond.)* 37, 255–260.
Bird, E.D., Chiappa, S.A. and Fink, G. (1976). *Nature* 260, 536–538.

Bird, E.D. and Iversen, L.L. (1974a). *Lancet* **i**, 463.
Bird, E.D. and Iversen, L.L. (1974b). *Brain* **97**, 457–472.
Bird, E.D., MacKay, A.V.P., Rayner, C.N. and Iversen, L.L. (1973). *Lancet* **i**, 1090–1092.
Caro, A.J. (1977a). M.D. Thesis, University of London.
Caro, A.J. (1977b). *Journal of the Royal College of Gen. Practitioners* **27**, 41–45.
Enna, S.J., Bennett, J.P. Jr., Bylund, D.B., Snyder, S.H., Bird, E.D. and Iversen, L.L. (1976). *Brain Res.* **116**, 531–537.
Hans, M.B. and Gilmore, T.H. (1969). *J. Nerv. Ment. Dis.* **148**, 5–13.
Hayden, M.R., Paul, M., Vinik, A.I. and Beighton, P. (1977). *Lancet* **ii**, 423–426.
Heathfield, K.W.G. and MacKenzie, I.C.K. (1971). *Guys Hospital Reports* **120**, 295–309.
Hornykiewicz, O. (1963). *Wien. Klin. Wschr.* **75**, 309–312.
Kanazawa, I. and Toyokura, Y. (1975). *Brain Res.* **100**, 371–381.
Merritt, A.D., Conneally, P.M., Rahman, N.F. and Drew, A.L. (1969). *In* "Progress in Neuro-genetics" (eds. Barbeau, A. and Brunette, J.-R.), Vol. I, pp. 645–650. Excerpta Medica Foundation, Amsterdam.
Perry, T.L., Hansen, S. and Kloster, M. (1973). *N. Engl. J. Med.* **288**, 337–342.
Phillipson, O.T. and Bird, E.D. (1977). *Clin. Science and Mol. Med.* **52**, 311–318.
Pleydell, M.J. (1954). *Brit. Med. J.* **ii**, 1121.
Podolsky, S. and Leopold, N.A. (1974). *J. of Clin. Endocrinol. and Metabol.* **39**, 36–39.
Reed, S.C. and Palm, J.D. (1951). *Science* **113**, 294–296.
Vessie, R.P. (1932). *J. Nerv. Ment. Dis.* **76**, 553–573.

SEROTONIN AND ACTION MYOCLONUS – A REVIEW

D. CHADWICK, M. HALLETT, P. JENNER and C.D. MARSDEN

University Department of Neurology,
King's College Hospital and the Institute of Psychiatry, London, U.K.

Introduction

Serotonin or 5-hydroxytryptamine (5HT) was first isolated from the central nervous system (CNS) of animals some 25 years ago (Twarog and Page, 1953). Although the metabolism of 5HT and the anatomical pathways of 5HT neurones are known, the functional role of this presumed neurotransmitter remains controversial. Hitherto, abnormalities of central 5HT metablism have not been associated with any specific neurological syndrome.

In 1971 Lhermitte and his colleagues reported that the administration of the 5HT precursor, L-5-hydroxytryptophan (5HTP) abolished myoclonus in a patient with post-anoxic encephalopathy (Lance and Adams, 1963). Several groups of workers have subsequently confirmed and extended these observations (Lhermitte *et al*, 1972; Guilleminault *et al*, 1973; Van Woert and Sethy, 1975; Chadwick *et al*, 1975; Van Woert *et al*, 1976 and 1977; Growdon *et al*, 1976; De Lean *et al*, 1976; Chadwick *et al*, 1977a). 5HTP may also be helpful in some other myoclonic syndromes (Chadwick *et al*, 1975; Van Woert *et al*, 1976; Growdon *et al*, 1976).

This report reviews our personal experience of the role of 5HT in myoclonus, and the experience of others described in the literature to date. It concentrates on a number of important questions. Firstly, which forms of myoclonus are responsive to 5HTP? Secondly, can biochemical and pharmacological analysis of patients with myoclonus help with the complex problem of classifying the many differing clinical and physiological phenomena previously called myoclonus? Thirdly, is there a basic biochemical disorder in patients with myoclonus responsive to 5HTP, and if so, what is it? Finally, what is the physiological basis of myoclonus in such patients?

The Clinical Features of Myoclonus Responsive to 5HTP

Two major problems exist in defining those forms of myoclonus responding to 5HTP, i.e. the difficulty in assessing improvement in myoclonic patients and the difficulty in classifying human myoclonus.

Most workers have adopted observer-scoring systems of the degree to which myoclonus impairs motor function. However, it is apparent that considerable variation in the severity of myoclonus can occur on a day-to-day basis, independent of any therapy, making it essential to employ placebo-control when assessing the effects of drug therapy on myoclonus. Even so, we have found it necessary to regard any improvement in myoclonus of less than 40% as of doubtful significance (Chadwick *et al*, 1977a). Some observers have reported smaller improvements as being 'significant', or have reported improvements in such terms as 'good', 'excellent', etc. The real significance of such reports is, therefore, difficult to assess.

The second difficulty arises because myoclonus is no more than a crude clinical description of a variety of rapid abnormal movements, which occur in many syndromes of differing pathology. Traditionally, the classification of myoclonus has been based on aetiology (Halliday 1967a). Such a classification is, however, difficult to apply, and it is doubtful whether different observers would agree entirely in the classification of myoclonus. Furthermore, there are a number of physiological types of myoclonus, and one physiological type may be produced by a number of pathological causes, whilst a single aetiological factor may result in myoclonus of differing physiologies in different patients (Halliday, 1967b; Chadwick *et al*, 1977a). Accordingly, it is necessary to consider the aetiological and physiological characteristics of myoclonus separately.

Table I summarises the accumulated aetiological data from eight publications. Post-anoxic and post-traumatic myoclonus responds most consistently to 5HTP, administered either alone or in combination with a peripheral decarboxylase inhibitor. 22 of 25 (88%) patients with post-anoxic myoclonus showed at least a moderate response, whilst all three patients with post-traumatic myoclonus responded well. As it is usually difficult to exclude a period of hypoxia following severe head injuries, it may be that post-anoxic and post-traumatic cases represent a single aetiological grouping.

Some patients with myoclonus of different aetiology also respond to treatment with 5HTP. Chadwick *et al* (1977a) report one patient with a non-progressive focal myoclonus of unknown aetiology, responding dramatically to 5HTP. Van Woert *et al* (1976) reported one patients with 'progressive myoclonic epilepsy' who showed a good response, and Growdon *et al* (1976) similarly report one

TABLE I

Myoclonic Syndromes and their Response to 5HTP

	Post-Anoxic Myoclonus	Post-Traumatic Myoclonus	Myoclonus & Cerebellar Disease (Ramsay Hunt Syndrome)	Lipidosis	Idiopathic Epilepsy & Myoclonus	Familial Familial Essential Myoclonus	Others
Lhermitte et al. (1971)	++ (1)						
Lhermitte et al. (1972)	++ (1)						
Guilleminault et al. (1973)	++ (1)		0 (3)				
Van Woert et al. (1976)	++ (4) + (5) 0 (3)	++ (1) + (1)	0 (1)			0 (2)	? Encephalitis 0 (1) 'Progressive myoclonic epilepsy' + (1)
De Lean et al. (1976)	++ (2)						
Growdon et al. (1976)	++ (2)			− (2)	− (1)		Unknown ++ (1) Congenital encephalopathy 0 (1) Unknown − (1)
Chadwick et al. (1977a)	++ (3) + (3)	++ (1)	0 (2)		0 (1)	0 (2)	Unknown ++ (1) ? Encephalitis 0 (1) Palatal myoclonus 0 (1)
Magnussen et al. (1977a)							Palatal myoclonus ++ (1)

++ Excellent response; + Good; 0 Unchanged; − Worse.
Figures in parenthesis represent number of cases.

patient with a six year 'progressive myoclonic syndrome of unknown aetiology'. Few clinical details are given concerning these latter two patients, so that it is difficult to decide to which sub-division of the 'progressive myoclonic epilepsy' (PME) syndrome they might belong. Halliday (1967a) describes three different groups of patients under the term PME; i.e. those with lipidoses; those with Lafora body disease; and those with multisystem disease as described by Ramsay Hunt (1921). All six patients who clearly had a Ramsay Hunt type of syndrome were unchanged by therapy with 5HTP (Table I), and both cases of lipid storage disease and myoclonus (Growden *et al*, 1976) became worse on 5HTP. It is, thus, tempting to speculate that those patients with PME responding to 5HTP might be examples of Lafora body disease. This possibility deserves further study.

To date, no patients with idiopathic epilepsy and myoclonus, or familial essential myoclonus, have shown convincing improvement with 5HTP. One report of improvement in palatal myoclonus following 5HTP has recently appeared (Magnussen *et al*, 1977), but it was ineffective in another case (Chadwick *et al*, 1977a). A beneficial effect of 5HTP on palatal myoclonus is difficult to explain. Most authorities regard it as very different from other forms of myoclonus. Halliday (1967a) suggests that it is more akin to nystagmus than myoclonus.

In all those subjects responsive to 5HTP who were described in sufficient detail, myoclonus was markedly exacerbated by voluntary movement and by intention, i.e., myoclonus was of the action type (Lance and Adams, 1963).

In many patients, the myoclonus could be evoked by a variety of sensory stimuli, including sudden noise, touch and passive movement. Such a relationship to sensory stimuli was not always clinically evident, but could be revealed by detailed physiological examination (Chadwick *et al*, 1977a). However, some patients with clinically similar action reflex myoclonus failed to respond to 5HTP (Growdon *et al*, 1976; Chadwick *et al*, 1977a). In such cases, action myoclonus occurred in association with multisystem diseases (Table I).

Those patients with myoclonus not of the action type *always* failed to respond to 5HTP (Chadwick *et al*, 1977a). The distribution of myoclonus has not been found helpful in predicting the response of myoclonus to 5HTP — both focal and generalised myoclonus being benefited.

Thus, it appears that patients with post-traumatic and post-anoxic myoclonus and some other patients with clinically similar forms of action reflex myoclonus respond to therapy with 5HTP. However, clinical criteria alone cannot predict the response to this drug with complete accuracy.

5HTP in Other Neurological Disorders

The effects of 5HTP are clearly specific to certain types of action myoclonus. It has been shown to be ineffective in controlling cerebellar tremor, essential tremor, dystonias, chorea, and tardive dyskinesias (Guilleminault *et al*, 1973; Chadwick *et al*, 1977a; Van Woert *et al*, 1977).

It has yet to be shown whether or not 5HTP is effective in controlling seizures. In those patients with action myoclonus and generalised convulsions, reported by Van Woert *et al* (1977), seizures became less frequent, or were abolished on 5HTP. It may be that this effect was due solely to reduction in the frequency of myoclonus, because three patients with idiopathic epilepsy unassociated with myoclonus, reported by the same observers, showed a slightly increased seizure-frequency whilst taking 5HTP. Furthermore, seizures became more frequent in two patients with lipidoses whose myoclonus also became worse (Growdon *et al*, 1976). In view of animal and clinical data suggesting a role for 5HT in the control of seizure threshold (Chadwick *et al*, 1977b) the anticonvulsant properties of 5HTP need to be fully evaluated.

The Physiology of 5HTP-Responsive Myoclonus

Chadwick *et al* (1977a) have studied several myoclonic patients in an attempt to characterise the physiology of myoclonus responsive to 5HTP. They found that patients in whom the EMG correlate of the myoclonic jerk was a prolonged period (150–300msec) of activity, failed to respond to 5HTP or clonazepam. In all those cases which did respond, the EMG correlate of the myoclonus was a brief (10–30msec) hypersynchronous action potential, similar to that seen during a monosynaptic spinal reflex.

Chadwick *et al* (1977a) tentatively identified two forms of post-anoxic and post-traumatic myoclonus, characterised by differing response to 5HTP and clonazepam. Those patients in whom the response to these agents was dramatic showed the following physiological characteristics:

   a) Proximal limb muscles were involved in the myoclonus.

   b) Sensory stimuli evoked a myoclonic response in all the muscles of the body (or in all those muscles which were capable of such a response in patients with focal myoclonus).

   c) The EEG showed spike/slow wave activity which was variably related to EMG activity.

   d) The somatosensory evoked potential (S.E.P's) were of normal amplitude.

   e) The cranial nerve musculature appeared to be activated in an ascending order, up the brain-stem.

They concluded that this form of myoclonus — reticular reflex myoclonus — was generated in the brain-stem (Hallett *et al*, 1977).

Other patients with post-anoxic myoclonus responding only moderately well to 5HTP and clonazepam differed in that:

   a) The myoclonus involved only distal limb muscles.

   b) Sensory stimuli evoked a myoclonic response affecting only the distal muscles of the limb which were stimulated.

   c) The EEG activity was without spike-wave activity but the myoclonic jerk appeared strictly related in time to a preceding cortical potential change.

   d) The SEP was enlarged.

   e) Brain-stem structures were activated in a downward direction.

It was therefore suggested that this myoclonus — cortical reflex myoclonus — originates in the cerebral cortex or hemispheres (Hallett *et al*, 1978).

These two forms of myoclonus are not mutually exclusive, for one patient with post-traumatic myoclonus exhibited the physiological characteristics of both types at different times.

The Pharmacology of 5HTP-Responsive Myoclonus

The administration of 5HTP to animals causes an elevation of brain 5HT and associated behavioural effects (Modigh, 1974). It is tempting to speculate that the effectiveness of 5HTP in suppressing myoclonus in man might be due to its effect on cerebral 5HT. Whilst there are many inconsistencies, the over-all pharmacological evidence is in keeping with such an hypothesis.

Agents other than 5HTP which increase brain concentrations of 5HT have a beneficial effect in 5HTP-responsive myoclonus — e.g., monoamine oxidase inhibitors, and L-tryptophan (the physiological precursor of 5HT) in combination with a monoamine oxidase inhibitor (Lhermitte *et al*, 1971 and 1972; Chadwick *et al*, 1975; De Lean *et al*, 1976) (see Table II).

Whether or not L-tryptophan given alone is therapeutically active is controversial. Van Woert and Sethy (1975) and Chadwick *et al* (1975 and 1977a) found no effect, whilst De Lean *et al* (1976) found it to be moderately effective. However, a lack of effect of L-tryptophan might not necessarily detract from the hypothesis that suppression of

TABLE II

Effects of Other Drugs Manipulating Central 5HT on Myoclonus Responsive to 5HTP

	Agents Elevating Brain 5HT			Agents Depleting 5HT	Presumed 5HT Receptor Antagonists
	Monoamine oxidase inhibitors (MAOI)	L-Tryptophan	L-Tryptophan + MAOI	p-Chlorophenylalanine	Methysergide
Lhermitte et al. (1971 & 1972)	+ (1)				− (2)
Romero et al. (1974)					− (1)
Bedard & Bouchard (1974)					+ (1)
Van Woert & Sethy (1975		0 (1) 0 (1)			0 (2) + (1)
De Lean et al. (1976)	+ (1)	+ (2)	+ (2)	0 (2)	+/− (1) 0/− (1)
Chadwick et al. (1977)	+ (4)	0 (4)	++ (4)		0 (1)

++ Excellent response
 + Good response
 0 No response
 − Worse
Figures in parenthesis indicate number of cases.

myoclonus is related to an increase in the quantities of functionally active cerebral 5HT. Animal evidence suggests that 5HT synthesised from L-tryptophan may fail to enter functionally active pools (Grahame-Smith, 1971).

The administration of the presumed 5HT antagonist, methysergide, has been reported to exacerbate myoclonus (Lhermitte *et al*, 1971 and 1972; Romero *et al*, 1974; De Lean *et al*, 1976), to have no effect (Van Woert *et al*, 1975; Chadwick *et al*, 1977a), or even occasionally to improve myoclonus (Bedard and Bouchard, 1974; De Lean *et al*, 1976). The studies of De Lean *et al* (1976) suggest that these varying results might to some degree be explained by a dose-variable effect of methysergide — causing an improvement at low doses, and an exacerbation at higher doses. Difficulties in interpreting the results with this drug arise from the fact that it crosses the blood brain barrier only poorly (Doepfner, 1962), and, most importantly, that it may possess both 5HT agonist (Martin and Eades, 1970) and antagonist properties (Klawans *et al*, 1973).

De Lean *et al*, (1976) administered para-chlorophenylalanine (pCPA), a relatively specific inhibitor of 5HT synthesis, to two patients with post-anoxic myoclonus. There was no exacerbation of the myoclonus in these patients, although CSF 5-hydroxyindole acetic acid (5HIAA), the 5HT metabolite, was reduced in one patient.

It is, however, difficult to know in what dose and for how long pCPA must be administered to man in order to produce a functionally significant depletion of brain 5HT.

Some workers have attempted to assess the effect of drugs manipulating central neurotransmitters, other than 5HT. The administration of L-dopa has been reported to suppress myoclonus in one patient with post-anoxic myoclonus (Minoli and Tredici, 1974) and to improve it in another (Lhermitte *et al*, 1972). Other workers have reported only minor improvement or deterioration with this agent (Van Woert and Sethy, 1975; De Lean *et al*, 1976). Drugs acting as dopamine receptor antagonists exacerbate myoclonus (Lhermitte *et al*, 1971 and 1972). The β-adrenoreceptor antagonist, propanolol, was without effect in one patient with post-anoxic myoclonus, but physostigmine resulted in a marked deterioration (Van Woert and Sethy, 1975). What is clear is that the degree of improvement produced by manipulation of transmitters other than 5HT is markedly less than that seen following 5HTP.

The anticonvulsant drug, sodium valproate, which appears to increase cerebral GABA concentrations, has been reported to be of great benefit in myoclonic syndromes associated with epilepsy, particularly those of childhood and adolescence (Lance and Anthony, 197 1977; Jeavons *et al*, 1977). However, the myoclonus in such syndromes may represent a very different phenomenon from post-anoxic myoclonus in the adult. Even so, it is of interest that valproate may also elevate levels of cerebral 5HT (Horton *et al*, 1977). Clearly the effect of sodium valproate on post-anoxic myoclonus needs to be investigated.

Of great interest is the beneficial effect of the benzodiazepine derivative, clonazepam, in post-anoxic myoclonus (Boudouresques *et al*, 1971; Chadwick *et al*, 1975; Goldberg and Dorman, 1976). It is as effective as 5HTP, in precisely those patients responsive to 5HTP. It has been shown to produce behavioural effects in animals which appear to be dependent on its ability to increase brain 5HT concentrations (Jenner *et al*, 1975; Chadwick *et al*, 1978). As clonazepam causes a significant elevation of CSF 5HIAA in humans, in doses that are therapeutically active against myoclonus, it may be that its effectiveness is dependent on its ability to alter brain 5HT metabolism (Chadwick *et al*, 1977a). The same may be true of the milder anti-myoclonic action of conventional anticonvulsants, as these also alter cerebral 5HT metabolism in the clinical situation (Chadwick *et al*, 1977b).

What then is the best practical drug regime at present for the treatment of action myoclonus? All workers have reported considerable gastro-intestinal side effects with 5HTP, even when combined with a peripheral decarboxylase inhibitor. It may also have unwanted central effects producing disturbances of mood (Van Woert *et al*, 1976; Chadwick *et al*, 1977a). Furthermore, 5HTP is not generally available. In spite of these difficulties, Van Woert and his co-workers have reported considerable long-term benefits in post-anoxic myoclonus following therapy with 5HTP in combination with a peripheral decarboxylase inhibitor (Van Woert *et al*, 1977).

The combination of L-tryptophan with a monoamine oxidase inhibitor provides a viable alternative, although its therapeutic effects may be rather less than those of 5HTP. Postural hypotension also may be a problem (Chadwick *et al*, 1977a). Thus the treatment of choice for post-anoxic myoclonus would currently appear to be clonazepam, which is usually well tolerated and as effective as 5HTP (Chadwick *et al*, 1977a).

CSF Biochemistry and 5HTP-Responsive Myoclonus

Lhermitte *et al* (1972) first reported a relatively low lumbar CSF concentration of 5HIAA in a patient with post-anoxic myoclonus responsive to 5HTP. This observation has been repeatedly confirmed (Guilleminault *et al*, 1973; De Lean *et al*, 1976; Chadwick *et al*, 1977a). Low CSF 5HIAA levels have also been found following the adminis-tration of probenecid to block the egress of 5HIAA from the CSF (Van Woert and Sethy, 1975; Van Woert *et al*, 1976 and 1977). Low CSF 5HIAA has also been reported in other myoclonic syndromes responsive to 5HTP, and Chadwick *et al* (1977a) found this to be of great value in predicting the response of myoclonus to 5HTP. These findings of low CSF 5HIAA suggest a reduced turnover of brain 5HT (Moir *et al*, 1970). That the reduction of CSF 5HIAA is of functional importance is indicated by the observation that the concentration of 5HIAA in the CSF rises with treatment with drugs causing improve-ment in the myoclonus. This is true not only following drug therapy, but also in one patient reported by Chadwick *et al* (1977a) whose myoclonus spontaneously resolved.

Chadwick *et al* (1977) also assayed CSF tryptophan, the physiological precursor of 5HT, and found this to be normal in patients with myoclonus. They did find low concentrations of CSF homovanillic acid (HVA), the dopamine metabolite, in 5HTP-responsive patients compared to control and to non-responsive myoclonic patients. Whilst this latter finding probably resulted from the inclusion of some patients with extrapyramidal deficits in the 5HTP-responsive group, Van Woert and Sethy (1975) also reported low levels of CSF HVA following probenicid administration in the patients with post-anoxic myoclonus. They did not, however, confirm this in a larger series of similar patients (Van Woert *et al*, 1976 and 1977). The significance of these findings, therefore, remains unknown.

The Biochemical Basis of Action Myoclonus

Myoclonic patients with low concentrations of CSF 5HIAA thus respond well to 5HTP and other agents causing a rise in CSF 5HIAA and by inference in cerebral 5HT. It therefore seems likely that 5HTP-responsive myoclonus is causally related to a cerebral deficiency of 5HT. What are the biochemical mechanisms which might produce a relative or absolute deficiency of 5HT and its metabolite?

The first possibility is that it could arise from a deficiency of L-tryptophan. The availability of this precursor may be important in controlling the rate of synthesis of 5HT (Wurtman and Fernstrom, 1972). This is, however, unlikely as CSF concentrations of tryptophan are normal in patients with myoclonus and L-tryptophan itself has doubtful therapeutic effects.

The second possibility is that low CSF concentrations of 5HIAA arise from a deficiency of tryptophan hydroxylase, as has been suggested by Van Woert and Sethy (1975). This would be in keeping with animal experimental evidence showing a decreased rate of tryptophan hydroxylation following cerebral anoxia (Davis and Carlsson, 1972), and with the crucial clinical observation that 5HTP, but not L-tryptophan alone, is effective in controlling myoclonus.

The conversion of L-tryptophan to 5HT (via 5HTP) requires an intact 5HT system, as tryptophan hydroxylase is found only within 5HT neurones (Kuhar *et al*, 1971). In contrast, aromatic acid decarboxylase, which convert 5HTP to 5HT, is present outside 5HT neurones (Corrodi *et al*, 1967), and the administration of high doses of 5HTP caused an abnormal distribution of 5HT (Moir and Eccleston, 1968), with idolamine fluorescence occurring inside dopaminergic and noradrenergic as well as serotoninergic neurones (Butcher *et al*, 1972). Therefore, if 5HT neurones were selectively destroyed during anoxia, it might be predicted that 5HT could be formed from 5HTP but not from L-tryptophan.

However, not all the facts are consistent with such a hypothesis. Firstly, Korf *et al* (1974) have shown that when 5HTP is administered to rats in low doses of 12.5 mg/kg (an amount which is greater than the doses administered in human studies) in combination with a peripheral decarboxylase inhibitor, the formation of 5HT may be dependent on a specific 5HTP decarboxylase related to 5HT neurones. Chadwick *et al* (1977) have also demonstrated that the administration of L-tryptophan is capable of elevating CSF 5HIAA (although not producing a therapeutic effect) in patients with 5HTP responsive myoclonus, and that the combination of L-tryptophan with a monoamine oxidase inhibitor has a therapeutic action. These facts suggest that an intact 5HT system with functioning tryptophan hydroxylase and 5HTP-decarboxylase is present in 5HTP-responsive myoclonus, and may indeed be necessary for the therapeutic action of 5HTP and L-tryptophan in combination with a monoamine oxidase inhibitor.

The finding that 5HTP, but not L-tryptophan itself, is effective in treating myoclonus probably reflects the fact that 5HT formed following L-tryptophan loading does not become functionally active (Grahame-Smith, 1971) in contrast to 5HT formed following 5HTP or L-tryptophan in combination with a monoamine oxidase inhibitor (Modigh, 1975).

A third possible explanation for the finding of low CSF 5HIAA must therefore be considered — i.e. that the 5HT neuronal system is hypoactive (but anatomically intact), possibly because of a loss of drive from other, as yet unidentified, damaged neuronal systems (Chadwick *et al*, 1977a). Further study is necessary to decide which of these two latter mechanisms is responsible for the low CSF 5HIAA found in patients with action myoclonus.

The Pathology of 5HTP-Responsive Myoclonus

Unfortunately, pathological data on patients with post-anoxic myoclonus are sparse. In the few cases so far coming to autopsy very diffuse anoxic damage was found (Castaigne *et al*, 1964; Wolf, 1977). It is impossible to localise the lesion responsible for myoclonus from such data. It would clearly be helpful to attempt to correlate full autopsy findings with regional biochemistry. Two of the cases reported by Chadwick *et al* (1977a) subsequently died and have been studied histologically and biochemically.

Both brains showed widespread anoxic damage, especially in cerebral cortex boundary zones and the hippocampus, but the basal

TABLE III

Concentrations of 5HT and 5HIAA in Cerebral Cortex and Striatum in Post-Anoxic Action Myoclonus+

| | 5 HT† (ug/g) | | | | 5 HIAA† (ug/g) | | | |
| | Cerebral Cortex | | Caudate Nucleus | Putamen | Cerebral Cortex | | Caudate Nucleus | Putamen |
	Area 8	Area 9			Area 8	Area 9		
Controls*	0.19 –	0.20 –	0.60 –	0.66 –	0.22 –	0.28 –	0.59 –	0.78 –
	0.70	0.61	0.87	0.69	0.38	0.40	0.72	1.32
Case A 73F	0.35	0.29	0.61	0.78	0.32	0.26	0.84	1.6
Case B 75M	0.28	0.24	0.39	0.50	0.23	0.27	0.41	0.55

+Neither of the two patients were taking drugs known to affect brain 5HT or 5HIAA in the week prior to death.
*The controls consisted of material from three subjects (68M, 73F, 80F) whose brains were removed at approximately the same time after death as those of the two patients, and which were assayed in parallel with the samples from the two patients.
†Assays carried out by Professor G. Curzon.

ganglia were not obviously affected, nor was there microscopic evidence of damage in the region of the mid-brain raphe region (Dr. I. Janota). Biochemical estimates of 5HT and 5HIAA in cerebral cortex and striatum of these two brains (Professor G. Curzon) are shown in Table III, in comparison to concurrent estimates in material from three control brains selected to match age, sex and time of brain removal after death. Neither cortical nor striatal 5HT and 5HIAA differed from control values in the two post-anoxic cases.

This limited evidence does not indicate any severe damage to 5HT systems in the brain, at least to those projecting to cerebral cortex and striatum. However, 5HT pathways to cerebellum and spinal cord have not been studied, and clearly more material is required before definitive conclusions can be reached.

Conclusions

The evidence reviewed here indicates that some forms of post-anoxic (and other) reflex-induced action myoclonus are associated with an apparent functional deficiency of brain 5HT. The administration of 5HT precursors and other drugs known to increase brain 5HT inhibits such reflex myoclonus, which appears to arise from abnormal activity of the brain stem reflex response to sensory stimuli. Such pathological and biochemical evidence as is available, however, does not point to an irreversible destruction of 5HT pathways projecting to cerebral cortex or the striatum, so the apparent 5HT deficiency in such cases may be due to functional inhibition of otherwise intact 5HT systems.

References

Bedard, P. and Bouchard, R. (1974). *Lancet* i, 738.

Boudouresques, J., Roger, J., Khalil, R., Vigouroux, R.A., Gosset, A., Pellissier, J.F. and Tassinari, C.A. (1971). *Rev. Neurol.* **125**, 306–309.

Butcher, L.L., Engel, J. and Fuxe, K. (1972). *Brain Res.* **41**, 387–411.

Castaigne, P., Cambier, J., Escourolle, R., Cathala, H.P. and Lecasble, R. (1964). *Rev. Neurol.* **111**, 60–73.

Chadwick, D., Harris, R., Jenner, P., Reynolds, E.H. and Marsden, C.D. (1975). *Lancet* ii, 434–435.

Chadwick, D., Hallett, M., Harris, R., Jenner, P., Reynolds, E.H. and Marsden, C.D. (1977a). *Brain* **100**, 455–481.

Chadwick, D., Jenner, P. and Reynolds, E.H. (1977b). *Ann. Neurol.* **1**, 218–224.

Chadwick, D., Gorrod, J.W., Jenner, P., Marsden, C.D. and Reynolds, E.H. (1978). *Brit. J. Pharmacol.* **62**, 115–124.

Corrodi, H., Fuxe, K. and Hokfelt, T. (1967). *J. Pharm. Pharmac.* **19**, 433–438.

Davis, J.N. and Carlsson, A. (1972). *J. Neurochem.* **20**, 913–915.

De Lean, J., Richardson, J.C. and Hornykiewicz, O. (1976). *Neurology (Minneap.)* **26**, 863–868.

Doepfner, W. (1962). *Experientia* **18**, 256–257.

Goldberg, M.A. and Dorman, J.D. (1976). *Neurology (Minneap.)* **26**, 24–26.

Grahame-Smith, D.G. (1971). *J. Neurochem.* **18**, 1053–1066.

Growdon, J.H., Young, R.R. and Shahani, B.T. (1976). *Neurology (Minneap.)* **26**, 1135–1140.

Guilleminault, C., Tharp, B.R. and Cousin, D. (1973). *J. Neurol. Sci.* **18**, 435–441.

Hallett, M., Chadwick, D., Adam, J. and Marsden, C.D. (1978). *J. Neurol. Neurosurg. Psychiat.* **40**, 253–264.

Hallett, M., Chadwick, D. and Marsden, C.D. (1978). Submitted to *Neurol. Minneap.*

Halliday, A.M. (1967a). *In* "Modern Trends in Neurology" (ed. Williams, D.), pp. 69–105. Butterworth, London.

Halliday, A.M. (1967a). *Brain* **90**, 243–284.

Horton, R.W., Anlezark, M., Sawaya, C.D. and Meldrum, B.S. (1977). *Eur. J. Pharmac.* **41**, 387–397.

Jeavons, P.M., Clark, J.E. and Maheshwari, M.C. (1977). *Dev. Med. Child Neurol.* **19**, 9–25.

Jenner, P., Chadwick, D., Reynolds, E.H. and Marsden, C.D. (1975). *J. Pharm. Pharmac.* **27**, 707–710.

Klawans, H.L., Goetz, B.A. and Weiner, W.J. (1973). *Neurol. (Minneap.)* **23**, 1234–1240.

Korf, J., Venema, K. and Postema, F. (1974). *J. Neurochem.* **23**, 249–252.

Kuhar, M.J., Roth, R.H. and Aghajanian, G.K. (1971). *Brain Res.* **35**, 167–176.

Lance, J.W. and Adams, R.D. (1963). *Brain* **86**, 111–136.

Lance, J.W. and Anthony, M. (1977). *Arch. Neurol.* **34**, 14–17.

Lhermitte, F., Marteau, R. and Degos, C.-F. (1972). *Rev. Neurol.* **126**, 107–114.

Lhermitte, F., Peterfalvi, M., Marteau, R., Gazengel, J. and Serdaru, M. (1971). *Rev. Neurol.* **124**, 21–31.

Magnussen, I., Dupont, E., Prange-Hansen, A. and Olivarius, B. de F. (1977). *Acta Neurol. Scand.* **55**, 251–253.

Martin, W.R. and Eades, C.G. (1970). *Psychopharmacologia* **17**, 242–257.

Minoli, G. and Tredici, G. (1974). *Lancet* ii, 472.

Modigh, K. (1974). *Acta Physiol. Scand. Suppl.* **403**, 1–56.

Moir, A.T.B., Ashcroft, G.W., Crawford, T.B.B., Eccleston, D. and Goldberg, H.G. (1970). *Brain* **93**, 357–368.

Moir, A.T.B. and Eccleston, D. (1968). *J. Neurochem.* **15**, 1093–1108.

Ramsay Hunt, J. (1921). *Brain* **44**, 490–538.

Romero, F., Gonzalez, F., Codina, A. and De Castro, J.L. (1975). *Lancet* i, 395–396.

Twarog, B.N. and Page, I.H. (1953). *Amer. J. Physiol.* **175**, 157–161.

Van Woert, M.H., Jutkowitz, R., Rosenbaum, D. and Bowers, M.B. Jr. (1976). *Monogr. neural. Sci.* **3**, 77–80.

Van Woert, M.H., Rosenbaum, D., Howieson, J. and Bowers, M.B. Jr. (1977). *New Eng. J. Med.* **296**, 70–75.

Van Woert, M.H. and Sethy, V.H. (1975). *Neurology (Minneap.)* **25**, 135–140.

Vogt, M. (1975). *In* "Proceedings of Sixth International Congress of Pharmacology" Vol. 2, pp. 3–9.

Wolf, P. (1977). *J. Neurol.* **215**, 39–47.

Wurtman, R.J. and Fernstrom, J.D. (1972). *In* "Perspectives in Neuropharmacology" pp. 143–193. Oxford University Press, New York.

NEUROTRANSMITTERS AND EPILEPSY

BRIAN MELDRUM

University Department of Neurology,
King's College Hospital and the Institute of Psychiatry, London, U.K.

All epileptic attacks are associated with the excessively synchronous or sustained discharge of a group of neurones. Inevitably this will be associated with changes in synaptic activity. It is therefore necessary to differentiate changes in synaptic function that are primary (i.e. that favour or precipitate seizures) from those that are secondary consequences of the abnormal neuronal discharges.

In seizures provoked by drugs or other toxic factors, there is often evidence for enhanced or diminished transmission at neurochemically-identified synapses. Among naturally occurring syndromes of epilepsy in animals and man some biochemical mechanisms involving transmitters have been identified, but for the major clinical syndromes the critical pathochemistry remains unknown.

Although many details are known about the actions of anticonvulsant drugs, an adequate account cannot yet be given of the mechanism by which seizures are prevented by any drug in clinical use.

In descriptions of mechanisms leading to epileptic discharges, and of mechanisms of action of anticonvulsant drugs, a distinction is often drawn between abnormalities or events concerning the nerve cell membrane and changes that are primarily in synaptic function. Obviously this cannot be an absolute distinction as key functional parts of the synapse (pre- and post-junction) are nerve cell membranes.

In vivo there is not always an evident difference between a drug acting at the receptor site and a drug acting in the membrane (to prevent, for example, an increase in chloride conductance following receptor site activation).

In what follows, mechanisms of epileptogenesis or of anticonvulsant action are discussed as they relate to the following proven or proposed central neurotransmitters: acetylcholine, glycine, GABA, dicarboxylic acids, dopamine, noradrenaline and serotonin.

Acetylcholine

The early literature on the possible involvement of central cholinergic mechanisms in epileptic phenomena was reviewed by Tower (1960). A comprehensive review of more recent work is provided by Maynert et al (1975).

Although the literature is replete with ambiguous and unconfirmed observations, the main points concerning acetylcholine may be simply summarised.

a) Cholinergic agonists (e.g. acetylcholine, nicotine and carbachol) applied experimentally to the cortical surface induce focal seizure discharges.

b) Compounds inhibiting acetylcholinesterase induce focal seizure activity when applied focally on the cortex, and generalised seizure activity when given systemically. This is true for reversible inhibitors such as physostigmine and for irreversible inhibitors such as the organophosphorus insecticides (e.g. diisopropylfluorophosphate).

c) Anticholinergics that enter the brain following systemic administration (e.g. atropine and scopolamine) rapidly terminate seizure activity induced by the agents described in a) and b). They do not prevent or terminate other forms of epilepsy in man.

Thus, pathologically augmented activity in a central cholinergic system can induce seizure activity. It appears likely that the system involved is the ascending cholinergic system delineated histochemically by Shute and Lewis (1967). This system participates in the initiation and maintenance of different phases of alertness. Enhanced activity (within the physiological range) is associated with activation of the EEG, and, in general, a reduced probability of occurrence of spike or spike and wave discharges. There is evidence that low doses of anticholinesterases can block epileptic discharges by a 'physiological' degree of enhancement of activity in this system (Williams and Russell, 1941).

Apart from poisoning with anticholinesterases, no epileptic syndrome in man has been shown to arise from either impaired or augmented central cholinergic function.

Although several in vitro effects of anticonvulsant drugs on cholinergic systems have been described, no class of anticonvulsants is believed to act primarily on cholinergic systems.

There is a genetically-determined epilepsy in EL mice (seizures induced by somatosensory stimulation, as provided by being repeatedly tossed in the air) which is possibly due to excessive activity in cholinergic systems. Such mice have an enhanced cerebral choline acetyltransferase activity, reduced cholinesterase activity and increased acetylcholine content (Kurokawa *et al*, 1966; Suzuki and Nakamoto, 1977).

It is possible that the myoclonus and generalised seizures induced experimentally by catechol (1,2-dihydroxybenzene) and some related compounds are due to enhanced activity in a cholinergic system (Angel *et al*, 1977).

Glycine

Glycine is an inhibitory transmitter released by some interneurones within the spinal cord (including Renshaw cells). The depressant action of glycine on neuronal firing is selectively blocked by strychnine, thebaine and related alkaloids (Curtis and Johnston, 1974). Convulsions induced by strychnine are characteristically spinal and it is clear that such seizures result from abolition of the normal inhibitory function of a glycinergic pathway in the spinal cord.

In human syndromes, the role of changed glycinergic activity in epileptic manifestations is less clear. Spasticity in experimental animals following ischaemic damage to the spinal cord is associated with a relatively selective loss of interneurones containing glycine (Davidoff *et al*, 1967) and a comparable pathology is probable in such spasticity in man. Curiously, however, scars and epileptogenic foci both in the human cortex and in the brains of experimental animals have an enhanced glycine content (Van Gelder *et al*, 1972). The functional meaning of this is unknown. Hyperglycinaemia, a congenital disorder of aminoacid metablism, is associated with muscular hypotonia. Generalised myoclonic jerks are observed and the EEG often shows spike and wave discharges. These epileptic manifestations are not necessarily a pharmacological consequence of the raised plasma concentration of glycine; they might, for example, arise from impaired formation of serine.

There is no clear evidence that any useful anticonvulsant drug acts by enhancing glycinergic transmission. Evidence that benzodiazepines compete for glycine binding sites in membrane preparations (Young *et al*, 1974) lacks the necessary support from physiological observations (Curtis *et al*, 1976a). Sodium valproate ('dipropylacetate') treatment can be associated with a marked elevation in plasma and urinary glycine concentration (Jaeken *et al*, 1977), but there is no accompanying elevation of CSF glycine content (unlike the situation in congenital non-ketotic hyperglycinaemia).

Gaba

That GABA acts as an inhibitory transmitter in many brain areas is now clearly established (see Ito, 1976). GABA is commonly synthesised in and released from small inhibitory interneurones, as for example, in the cerebral cortex and the hippocampus (Storm-Mathisen, 1977). These interneurones are activated by local (recurrent) pathways and appear to be part of the system that normally prevents the excessively synchronous and sustained firing of the neuronal masses. The malfunction of GABA-mediated transmission is readily envisaged as leading to epileptic discharges.

GAD Inhibition

Seizures can be produced experimentally by drugs that block the synthesis of GABA by inhibiting the enzyme glutamic acid decarboxylase (GAD). Unfortunately, the relationship between GAD inhibition and seizure induction became clouded in controversy because most of the initial studies were performed with hydrazides. These compounds have a wide variety of actions. Among other effects they impair the synthesis and coenzymic function of pyridoxal phosphate, which is an essential co-factor for all cerebral decarboxylases and transaminases, including GABA transaminase. Because of disparate interactions between these factors, there is not a good correlation between seizure incidence and either GAD inhibition or whole brain GABA content after different hydrazides (see Medina, 1963; Meldrum, 1975; Meldrum *et al*, 1975).

However, the correlation between GAD inhibition and seizure induction is much clearer for some other convulsant drugs. These can be classified into three categories (see Table I). Seizures following the simple competitive inhibitor, 3-mercaptopropionic acid, occur after a very short latency in mice and baboons, and the EEG features are similar to those seen after picrotoxin (Horton and Meldrum, 1973; Meldrum and Horton, 1974). Drugs that act as pyridoxal phosphate antagonists resemble allylglycine and its metabolite, 2-keto-4-pentenoic acid (which irreversibly inhibits the enzyme) in producing brief repetitive seizures after a long latency (Meldrum *et al*, 1970; Horton and Meldrum, 1973, 1977).

Regional differences in GAD inhibition, leading to local impairment of GABA synthesis, probably explain some of the differing seizure patterns seen after allylglycine or hydrazides (Horton *et al*, 1978).

The suggestion that inhibition of L-aromatic aminoacid decarboxylase makes an important contribution to seizure induction by thiosemicarbazide and other convulsant drugs (Gey and Georgi, 1974; Baxter, 1976) was not confirmed in a recent study (Sawaya

TABLE I

Convulsant Drugs that Inhibit Glutamic Acid Decarboxylase

	Enzyme inhibition	Convulsant Dose mmol/kg		Seizure latency (min)
		rodents	baboons	
3-mercaptopropionic acid	competitive	1.1	0.3 − 0.8	4−6
4-deoxypyridoxine	)	1.2	0.5 − 0.9	15−45
methyl-dithiocarbazinate	) pyridoxal	0.1	−	30−45
isoniazid	) phosphate) antagonists	−	0.7 − 1.0	7.40
thiosemicarbazide	)	0.3	0.04 − 0.1	60−240
allylglycine	) irreversible	1	4	60−120
2-keto-4-pentenoic acid	) inhibitors	(10−20ug)*	−	30−60

*Focal intracerebral injections inducing seizures in mice or rats.
Data from: Meldrum *et al*, 1970; Meldrum and Horton, 1971; Horton and Meldrum, 1973; Meldrum *et al*, 1975; Horton and Meldrum, 1977.

et al, 1977).

GABA Antagonists

Several convulsant drugs have been shown to block the inhibitory action of GABA on neurones. Some of these are listed in Table II. Picrotoxin and bicuculline share the capacity to prevent the inhibition of cortical neurones induced by iontophoresis of GABA; they also block the GABA-induced depolarisation of the rat superior cervical ganglion, but only bicuculline competes with radioactively labelled GABA for binding with high affinity sites in cerebral membrane preparations (Zukin *et al*, 1974; Enna *et al*, 1977). A possible explanation is that bicuculline (and the bicyclic phosphate esters) competes with GABA for binding at its receptor site, whereas picrotoxin interferes with the next event in the inhibitory action of GABA, by preventing the change in the ionophore that leads to stabilisation or enhancement of the resting potential.

All the GABA antagonists produce generalised, long-lasting seizures after an extremely or moderately short latency (Meldrum and Horton, 1974; Meldrum, 1975). The brain region in which they act to trigger generalised seizures when given systemically has not been defined. It appears to be subcortical in the case of bicuculline.

GABA Transaminase Inhibitors

GABA-transaminase (4-aminobutyrate : 2-oxoglutarate aminotrans-

TABLE II

Convulsant Drugs that are GABA Antagonists

	GABA Antagonism Demonstrated				
	ionto-phoretic-ally	on rat sup.cerv. ganglion	by com-petition for binding sites	convulsant dose in baboon mg/kg	seizure latency (min)
picrotoxin	++	++	0	1.5	2−10
bicuculline	++	++	++	0.3	0.1−0.2
4-isopropyl bicyclophosphate	++	++	+	0.05	1−2
tetramethylene-disulphotetramine		++	0	0.3	0.1−0.2

Data from Enna *et al*, 1977; Zukin *et al*, 1974; Bowery *et al*, 1976; Meldrum and Horton, 1974; Meldrum unpublished.

ferase, E.C. 2.6.1.19), is the first enzyme in the further metabolism of GABA. It is a mitochondrial enzyme (Waksman *et al*, 1968) that requires pyridoxal phosphate and has been obtained in a pure form (Schousboe *et al*, 1973). Three drugs that have been shown to be potent inhibitors of GABA-transaminase, amino-oxyacetic acid, hydrazinopropionic acid, and 3-mercaptopropionic acid, all produce convulsions in animal models (Van Gelder, 1969; Meldrum *et al*, 1970; Horton and Meldrum, 1973; Wu and Roberts, 1974). This is probably because all three drugs also potently inhibit GAD. Recently four catalytic inhibitors of GABA-transaminase, that combine with the active site of the enzyme and irreversibly inactivate it, have been described. Ethanolamine-O-sulphate, when administered intra-cerebroventricularly to mice, leads to a 4−10 fold increase in brain GABA content and blocks sound-induced seizures in suceptible strains (Fowler, 1973; Anlezark *et al*, 1976; Horton *et al*, 1977). γ-acetylenic-GABA and γ-vinyl-GABA produce similar effects when administered systemically in mice (Schechter *et al*, 1977a, b). They have also been shown to block photically-induced epilepsy in *Papio papio* when given intravenously (Meldrum and Horton, 1978). Gabaculine (Rando and Bangarter, 1976) is a naturally occurring catalytic inhibitor of GABA-transaminase that also raises brain GABA content and blocks some drug-induced seizures (Matsui and Deguchi, 1977). Sodium valproate (di-n-propylacetate) is a weak inhibitor of GABA-transaminase (Godin *et al*, 1969). The possibility that it owes its anticonvulsant action to this effect is discussed below.

GABA Agonists

GABA analogues that mimic the inhibitory action of GABA on neurone firing have long been expected to be anticonvulsant, but a useful anti-epileptic drug that is physiologically proven as a GABA-agonist has yet to be identified (see review by Meldrum, 1975).

Simple straight-chain analogues of GABA with double or triple bonds between the second and third carbon atoms (4-aminotetrolic acid and trans-4-aminocrotonic acid) mimic GABA in iontophoretic studies (Beart *et al*, 1971; Johnston *et al*, 1975) but have not been evaluated as anticonvulsants. Muscimol and isoguvacine are somewhat similar analogues, but with three of the carbons in the chain forming part of a ring structure. Their GABA-mimetic potency, assessed iontophoretically (Krogsgaard-Larson *et al*, 1977) or *in vitro* (Bowery *et al*, 1978) is clear, but their anticonvulsant action remains uncertain.

Two analogues that in iontophoretic studies have non-specific depressant properties (Curtis *et al*, 1974) produce sustained spike and wave discharges, resembling the EEG picture of petit mal, when given systemically to animals. These are the antispasticity drug baclofen, β-(parachlorophenyl) γ-aminobutyric acid (Meldrum and Horton, 1974); and the anaesthetic compound butyrolactone (or its hydrolysis product, 4-hydroxybutyric acid) (Winters and Spooner, 1965; Godschalk *et al*, 1976).

GABA Uptake Blockers

Several compounds that block reuptake of GABA into neurones or glia (such as nipecotic acid and 2,4-diaminobutyric acid) have been shown to prolong the inhibitory action of iontophoretically applied GABA (Curtis *et al*, 1976b). Their effects on epileptic processes have not yet been adequately evaluated.

Presynaptic Inhibition

The exact basis for the effects on transmission at the first afferent synapse that are sometimes referred to as 'presynaptic inhibition' is not certain, but it is associated with a depolarisation of the terminals that is detectable in the spinal cord as dorsal root potentials. Much indirect evidence suggests that GABA plays a role in this form of inhibition. Picrotoxin and bicuculline both diminish presynaptic inhibition (assessed by monosynaptic reflex reduction) and dorsal root potentials (Eccles *et al*, 1963; Barker and Nicoll, 1972). So, interestingly, do some convulsants not known to act on GABA mediated inhibition, such as pentylenetetrazol and bemegride (Hayes *et al*, 1977).

Presynaptic inhibition is enhanced and prolonged by barbiturates, benzodiazepines, trimethadione, and general anaesthetics such as chloroform and ether (Schmidt, 1971; Miyahara *et al*, 1966).

Anticonvulsant Drugs

Increases in the cerebral GABA content following systemic administration of hydantoins and barbiturates in rodents have been reported (Woodbury, 1969; Saad *et al*, 1972) but are not large, not consistent, and not correlated with anticonvulsant effect (Horton *et al*, 1976; Vial *et al*, 1974).

Barbiturates potentiate the actions of GABA on neurones in culture (Ransom and Barker, 1976), on the rat superior cervical ganglion (Evans, 1977), and on dorsal horn interneurones (Curtis and Lodge, 1977).

Various indirect arguments that benzodiazepines enhance GABA-mediated inhibition have been presented (Fuxe *et al*, 1975; Haefeley *et al*, 1975). The enhancement of dorsal root potentials after benzodiazepines (and barbiturates) is consistent with such a mechanism. Radioactive binding studies utilizing labelled diazepam indicate a site of action separate from that of GABA (Mohler and Okada, 1977a, b; Braestrup *et al*, 1977) and of the barbiturates.

In vivo studies showing regional differences in the changes in GABA metabolism after diazepam and pentobarbital support the concept that both drugs facilitate GABA-mediated inhibition, but at different sites (Pericic *et al*, 1977).

Sodium valproate given acutely in rather high doses to rats or mice elevates brain GABA content by 10–60% (Godin *et al*, 1969; Anlezark *et al*, 1976; Horton *et al*, 1977). This change is quantitatively greater than that reported after barbiturates or hydantoin, but is very much less than that associated with the anticonvulsant action of catalytic inhibitors of GABA-transaminase (see above). An inhibitory action of valproate (and of some other anticonvulsant drugs) on succinic semialdehyde dehydrogenase, the enzyme responsible for the last step in the GABA shunt, has also been described (Harvey *et al*, 1975; Sawaya *et al*, 1975). Enhancement of GABA-mediated inhibition may well be the mechanism of action of high doses of valproate in animals; the exact basis for the antiepileptic action of lower, chronic doses in man is still not clearly established.

Human Epilepsy and GABA

The only epileptic seizures in man that can, at present, confidently be ascribed to a failure of GABA-mediated inhibition are those occurring as a result of pyridoxine deficiency in infants (Coursin, 1964), and, probably, those related to poisoning with pyridoxal phosphate antagonists.

Dicarboxylic Acids (Glutamic and Aspartic)

The intracortical or intracerebroventricular injection of aspartate, or glutamate, leads to seizures (Hoyashi, 1954; Crawford, 1963).

If applied iontophoretically glutamate and aspartate are each excitatory on most central neurones (see Curtis and Johnston, 1974), but precise evidence of a physiological role still is lacking. Similarly, an exact role for these acids in epileptogenesis or anticonvulsant drug action has yet to be defined.

Some convulsant drugs interfere with the metabolism of glutamate (e.g. methionine sulphoximine inhibits glutamine synthetase, the enzyme 'inactivating' glutamate with ammonia; and 3-mercaptopropionic acid competes with glutamate for the active site of glutamate decarboxylase), but physiological inactivation of synaptically released glutamate is almost certainly through specific uptake systems into nerves and glia. Anticonvulsant activity is shown by a drug (1-hydroxy-3-amino-pyrrolid-2-one, HA966) that blocks the excitatory activity of glutamic acid (Bonta *et al*, 1971; Davies and Watkins, 1973).

Measurements of glutamic acid concentration in human or experimental cortical epileptogenic lesions indicate a reduced tissue concentration of glutamate and aspartate (Koyama, 1972; Van Gelder *et al*, 1972).

The corresponding sulphur-containing amino acids, cysteic and homocysteic acids, are also excitant when applied to spinal cord neurones, with a potency equal to or greater than that of aspartate and glutamate (Curtis *et al*, 1960; Biscoe *et al*, 1976). Homocysteine is convulsant when administered parenterally to rodents (Sprince *et al*, 1969). Because of the impaired activity of cystathionine synthetase, homocysteine accumulates in the brains of children with homocystinuria, in which condition generalised seizures are not uncommon. However, homocysteine is a pyridoxal phosphate antagonist (Taberner *et al*, 1977) and this may explain its convulsant effect (see above).

Dopamine and Noradrenaline

Although it is evident that cerebral catecholamines participate in the control of movement, their role in epileptic phenomena is not yet clear.

Drugs which deplete brain catecholamines lower convulsive

thresholds in animals (Maynert *et al*, 1969; Jobe *et al*, 1973). Conversely, drugs which enhance catecholaminergic transmission raise convulsive thresholds in some test systems, e.g. photosensitive baboons (Meldrum *et al*, 1975b). The change in seizure threshold observed in the 'kindling' process is associated with catecholamine depletion in the amygdala (Engel and Sharpless, 1977). However, there is no evidence that an abnormality of function in a catecholaminergic pathway is a primary cause of epilepsy in man.

Although a dependence of anticonvulsant drug action on normal monaminergic function has been shown in some animal models (Meyer and Frey, 1973), the importance of monoaminergic activity in anticonvulsant drug action in man remains uncertain. Possibly some of the movement disorders associated with diphenylhydantoin intoxication arise from blockade of dopaminergic transmission (Ahmad *et al*, 1975).

Serotonin

As with the catecholamines, depletion of cerebral serotonin lowers convulsive threshold, and enhancement of serotoninergic transmission raises seizure threshold in numerous animal test systems (see Meldrum *et al*, 1975c). In baboons with photosensitive epilepsy, administration of the immediate precursor, L-5-hydroxytryptophan, L-5HTP, 10–35 mg/kg, completely blocks myoclonic responses to photic stimulation (Wada *et al*, 1972). Several compounds known to produce serotonin-like actions at central sites, such as dimethyltryptamine and lysergic acid diethylamide, also block myoclonic responses (Meldrum and Naquet, 1971).

In man, several forms of reflex or intention myoclonus (most notably post-hypoxic action myoclonus, or Lance Adams syndrome) respond to the administration of L-5HTP, with or without a peripheral decarboxylase inhibitor (Lhermitte *et al*, 1972; Van Woert *et al*, 1977). The concentration of the serotonin metabolite, 5-hydroxy-indoleacetic acid, 5-HIAA, is reduced in the spinal CSF of patients with intention myoclonus (Van Woert *et al*, 1977). Patients with other forms of epilepsy show normal CSF 5-HIAA levels (Chadwick *et al*, 1977) and are not known to respond to L-5HTP therapy (see Growdon *et al*, 1976).

It has been suggested that seizures seen in patients after overdosage with tricyclic antidepressant drugs result from enhanced activity at serotoninergic synapses (Westheimer and Klawans, 1974). This seems unlikely as similar seizures induced by imipramine in baboons can be totally prevented by pretreatment with L-5HTP

(Trimble *et al*, 1977).

Anticonvulsant drugs administered in high doses to rodents lead to increases in the cerebral content of serotonin and 5-HIAA (Bonnycastle *et al*, 1957; Jenner *et al*, 1975; Horton *et al*, 1977). Anticonvulsant action, however, is seen in the absence of this effect, i.e. with lower doses or when 5-HT synthesis is impaired (Horton *et al*, 1977). In man, toxic doses of anticonvulsant drugs are associated with an increase in CSF 5-HIAA concentration (Chadwick *et al*, 1977). It is not certain whether 5HT turnover is increased or decreased by anticonvulsant drugs. The transport of acid metabolites (such as 5-HIAA) from CSF to blood is slowed by benzodiazepines and hydantoin (Chase *et al*, 1970) and this may explain some of the increase in brain and CSF 5HIAA content.

Electroconvulsive Therapy

Numerous changes in transmitter concentration in the rodent brain have been described during or a few hours after electroconvulsive shock (for serotonin see Essman, 1973) but their significance is uncertain.

Of greater possible relevance to the therapeutic action of ECT is the observation that behavioural responses to precursors or agonists of cerebral monoamines are enhanced in rats and mice that have received electroconvulsive shock daily for seven to ten days (Modigh, 1975; Evans *et al*, 1976). An enhanced sensitivity of post-synaptic receptors to monoamines would be consistent with current monoaminergic theories of the pathophysiology of depressive illness.

Conclusion

Changes in activity at specific synapses clearly can induce or prevent epileptic attacks. However, we know too little about cerebral neurotransmitters and about the pathophysiology of epilepsy to give an account of what is wrong at the synaptic level in the major syndromes of epilepsy, or how anticonvulsant drugs may put it right. We do not know which compounds, other than those described above, may be important. Substance P and several newly-identified peptides appear to act as neurotransmitters in the brain. Their functional significance is not yet clear, but already changes in epileptic activity produced by enkephalins have been reported (Urca *et al*, 1977).

Progress is hampered by the difficulty of determining what is happening at the synaptic level in the human brain. CSF studies are of limited value. Post-mortem studies may be helpful for chronic syndromes in the adult, but are unlikely to be available for the study of *petit mal*, 'benign' febrile convulsions, or human photosensitive epilepsy.

Further basic studies on cerebral neurotransmitters are essential before we can gain a detailed understanding of epileptic processes.

References

Ahmad, S., Laidlaw, J., Houghton, G.W. and Richens, A. (1975). *J. Neurol. Neurosurg. Psychiat.* **38**, 225–231.

Angel, A., Clarke, K.A., and Dewhurst, D.G. (1977). *Brit. J. Pharmacol.* **61**, 433–439.

Anlezark, G., Horton, R.W., Meldrum, B.S. and Sawaya, M.C.B. (1976). *Biochem. Pharmacol.* **25**, 413–417.

Barker, J.L. and Nicoll, R.A. (1972). *Science* **176**, 1043–1045.

Baxter, C.F. (1976). *In* "GABA In Nervous System Function" (eds. Roberts, E., Chase, T.N. and Towers, D.B.), pp. 61–87. Raven Press, New York.

Beart, P.M., Curtis, D.R. and Johnston, G.A.R. (1971). *Nature (New Biol.)* **234**, 80–81.

Biscoe, T.J., Evans, R.H., Headley, P.M., Martin, M.R. and Watkins, J.C. (1976). *Brit. J. Pharmacol.* **58**, 373–382.

Bonnycastle, D.D., Giarman, N.J. and Paasonen, M.K. (1957). *Brit. J. Pharmac.* **12**, 228–231.

Bonta, I.L., De Vos., C.J., Grijsen, H., Hiller, F.L., Nooch, E.L. and Sim, A.W. (1971). *Brit. J. Pharmacol.* **43**, 514–535.

Bowery, N.G., Collins, J.F. and Hill, R.G. (1976). *Nature* **261**, 601–603.

Bowery, N., Collins, J.F., Hudson, A.L. and Neal, M.J. (1978). *Experienta*, in press.

Braestrup, C., Albrechsten, R. and Squires, R.F. (1977). *Nature* **269**, 702–704.

Chadwick, D., Jenner, P. and Reynolds, E.H. (1977). *Ann. Neurol.* **1**, 218–224.

Chase, T.N., Katz, R.I. and Kopin, I.J. (1970). *Neuropharmacol.* **9**, 103–108.

Coursin, D.B. (1964). *Vitam. Horm. (New York)* **22**, 755–783.

Crawford, J.M. (1963). *Biochem. Pharmacol.* **12**, 1443–1444.

Curtis, D.R., Phillis, J.W. and Watkins, J.C. (1960). *J. Physiol.* **150**, 656–682.

Curtis, D.R., Game, C.J.A., Johnston, G.A.R. and McCulloch, R.M. (1974). *Brain Research* **70**, 493–499.

Curtis, D.R. and Johnston, G.A.R. (1974). *Ergbn. der Physiol.* **69**, 97–188.

Curtis, D.R., Game, C.J.A. and Lodge, D. (1976a). *Brit. J. Pharmac.* **56**, 307–311.

Curtis, D.R., Game, C.J.A. and Lodge, D. (1976b). *Exp. Brain Research* **25**, 413–428.

Curtis, D.P. and Lodge, D. (1977). *J. Physiol. (Lond.)* **272**, 48–49P.

Davidoff, R.A., Graham, L.T., Shank, R.P., Werman, R. and Aprison, M.H. (1967). *J. Neurochem.* **14**, 1025–1031.

Davies, J. and Watkins, J.C. (1973). *Neuropharmacol.* **12**, 637–640.

Eccles, J.C., Schmidt, R. and Willis, W.D. (1963). *J. Physiol. (Lond.)* **168**, 500–530.

Engel, J. and Sharpless, N.S. (1977). *Brain Research* **136**, 381–386.

Enna, S.J., Collins, J.F. and Snyder, S.H. (1977). *Brain Research* **124**, 185–190.

Essman, W.B. (1973). Neurochemistry of cerebral electroshock. Spectrum Publications, New York.

Evans, R.H. (1977). *J. Physiol. (Lond.)* **272**, 49–50P.

Evans, J.P.M., Grahame-Smith, D.G., Green, A.R. and Tordoff, A.F.C. (1976). *Brit. J. Pharmac.* **56**, 193–199.

Fowler, L.J. (1973). *J. Neurochem.* **21**, 437–440.

Fuxe, K., Agnati, L.F., Bolme, P., Hokfelt, T., Lidbrink, P., Ljungdahl, A., Perez de la Mora, M. and Ogren, S.O. (1975). *In* "Mechanism of Action of Benzodiazepines" (eds. Costa, E. and Greengard, P.) pp. 45–61. Raven Press, New York.

Gey, K.F. and Georgi, H. (1974). *J. Neurochem.* **23**, 725–738.

Godin, Y., Heiner, L., Mark, J. and Mandel, P. (1969). *J. Neurochem.* **16**, 869–873.

Godschalk, M., Dzoljic, M.R. and Bonta, I.L. (1976). *Neuroscience Letters* **3**, 145–150.

Growdon, J.H., Young, R.R. and Shahani, B.T. (1976). *Neurology (Minneap.)* **26**, 1135–1140.

Haefely, W., Kulcsar, A., Mohler, H., Pieri, L., Polc, P. and Schaffner, R. (1975). *In* "Mechanism of Action of Benzodiazepines" (eds. Costa, E. and Greengard, P.), pp. 131–151. Raven Press, New York.

Harvey, P.K.P., Bradford, H.F. and Davison, A.N. (1975). *Febs Letters* **52**, 251–253.

Hayashi, T. (1954). *Keio J. Med.* **3**, 183–192.

Hayes, A.G., Gartside, I.B. and Straughan, D.W. (1977). *Neuropharmacol.* **16**, 725–730.

Horton, R.W. and Meldrum, B.S. (1973). *Brit. J. Pharmacol.* **49**, 52–63.

Horton, R.W. and Meldrum, B.S. (1977). *Brit. J. Pharmacol.* **61**, 477P.

Horton, R.W., Meldrum, B.S., Sawaya, M.C.B. and Stephenson, J.D. (1976). *Eur. J. Pharmacol.* **40**, 101–106.

Horton, R.W., Anlezark, G.L., Sawaya, M.C.B. and Meldrum, B.S. (1977). *Europ. J. Pharmacol.* **41**, 387–397.

Ito, M. (1976). *In* "GABA in Nervous System Function" (eds. Roberts, E., Chase, T.N. and Tower, D.B.), pp. 427–448. Raven Press, New York.

Jaeken, J., Corbeel, L., Casaer, P., Carchon, H., Eggermont, E. and Eekels, R. (1977). *Lancet* **II**, 617.

Jenner, P., Chadwick, D., Reynolds, E.H. and Marsden, C.D. (1975). *J. Pharm. Pharmac.* **27**, 707–710.

Jobe, P.C., Picchioni, A.L. and Chin, L. (1973). *J. Pharmac. exp. Therap.* **184**, 1–10.

Johnston, G.A.R., Curtis, D.R., Beart, P.M., Game, C.J.A., McCulloch, R.M. and Twitchin, B. (1975). *J. Neurochem.* **24**, 157–160.

Koyama, I. (1972). *Canad. J. Physiol. Pharmacol.* **50**, 740–752.

Krogsgaard-Larsen, P., Johnston, G.A.R., Lodge, D. and Curtis, D.R. (1977). *Nature* **268**, 53–55.

Kurokawa, M., Naruse, H. and Kato, M. (1966). *In* "Progress in Brain Research", **21A**, 112–130.

Lhermitte, F., Marteau, R. and Degos, C.F. (1972). *Rev. Neurol.* **126**, 107–114.

Matsui, Y. and Deguchi, T. (1977). *Life Sciences* **20**, 1291–1296.

Maynert, E.W., Marczynski, T.J. and Browning, R.A. (1975). *In* "Advances in Neurology", Vol. 13, pp. 79–147. Raven Press, New York.

Maynert, E.W. (1969). *Epilepsia* **10**, 145–162.

Medina, M.A. (1963). *J. Pharmacol. exp. Therap.* **140**, 133–137.

Meldrum, B.S. (1975). *Internat. Rev. Neurobiol.* **17**, 1–36.

Meldrum, B.S. and Horton, R.W. (1971). *Brain Research* **35**, 419–436.

Meldrum, B.S. and Horton, R.W. (1974). *In* "The Natural History and Management of Epilepsy" (eds. Harris, P. and Mawdsley, C.), pp. 55–64. Churchill Livingstone, Edinburgh.

Meldrum, B.S. and Horton, R.W. (1978). *Psychopharmacology* (in press).

Meldrum, B.S. and Naquet, R. (1971). *Electroenceph. clin. Neurophysiol.* **31**, 563–572.

Meldrum, B.S., Balzano, E., Gadea, M. and Naquet, R. (1970). *Electroenceph. clin. Neurophysiol.* **29**, 333–347.

Meldrum, B.S., Horton, R.W. and Sawaya, M.C.B. (1975a). *J. Neurochem.* **24**, 1003–1010.

Meldrum, B.S., Anlezark, G. and Trimble, M. (1975b). *Europ. J. Pharmacol.* **32**, 203–213.

Meldrum, B.S., Anlezark, G., Balzamo, E., Horton, R.W. and Trimble, M. (1975c). *In* "Advances in Neurology" (eds. Meldrum, B.S. and Marsden, C.D.), Vol. 10, pp. 119–128. Raven Press, New York.

Meyer, H. and Frey, H.H. (1973). *Neuropharmacol.* **12**, 939–947.

Miyahara, J.T., Esplin, D.W. and Zablocka, B. (1966). *J. Pharmacol. exp. Therap.* **154**, 118–127.

Modigh, K. (1975). *J. Neural Transmission* **36**, 19–32.

Mohler, H. and Okada, T. (1977). *Life Sciences* **20**, 2101–2110.

Mohler, H. and Okada, T. (1977). *Science* **198**, 849–851.

Pericic, D., Walters, J.R. and Chase, T.N. (1977). *J. Neurochem.* **29**, 839–846.

Rando, R.R. and Bangerter, F.W. (1976). *J. Amer. Chem. Soc.* **98**, 6762–6764.

Ransom, B.R. and Barker, J.L. (1976). *Brain Res.* **114**, 530–535.

Saad, S.F., El Masry, A.M. and Scott, P.M. (1972). *Eur. J. Pharmacol.* **17**, 386–392.

Sawaya, M.C.B., Horton, R.W. and Meldrum, B.S. (1975). *Epilepsia* **16**, 649–655.

Sawaya, C., Horton, R. and Meldrum, B. (1977). *Biochem. Pharmacol.* (in press).

Schechter, P.J., Tranier, Y., Jung, M.J. and Sjoerdsma, A. (1977a). *J. Pharmacol. exp. Therap.* **201**, 606–612.

Schechter, P.J., Tranier, Y., Jung, M.J. and Bohlen, P. (1977b). *Europ. J. Pharmacol.* **45**, 319–328.

Schmidt, R.F. (1971). *Ergeb. Physiol.* **63**, 20–101.

Schousboe, A., Wu, J.Y. and Roberts, E. (1973). *Biochemistry* **12**, 2868–2873.

Shute, C.C.D. and Lewis, P.R. (1967). *Brain* **90**, 497–520.

Sprince, H., Parker, C.M., Josephs, J.A. and Magazino, J. (1969). *An. N.Y. Acad. Sci.* **166**, 323–325.

Storm-Mathisen, J. (1977). *Progress in Neurobiol.* **8**, 119–181.

Suzuki, J. and Nakamoto, Y. (1977). *Electroenceph. clin. Neurophysiol.* **43**, 299–311.

Taberner, P.V., Pearce, M.J. and Watkins, J.C. (1977). *Biochem. Pharmacol.* **26**, 345–349.

Tower, D.B. (1960). Neurochemistry of Epilepsy. C.C. Thomas, Springfield, Ill.

Trimble, M., Anlezark, G. and Meldrum, B. (1977). *Psychopharmacology* **51**, 159–164.

Urca, G., Frenk, H., Liebeskind, J.C. and Taylor, A.N. (1977). *Science* **197**, 83–86.

Van Gelder, N.M. (1969). *J. Neurochem.* **16**, 1355–1360.

Van Gelder, N.M., Sherwin, A.L. and Rasmussen, T. (1972). *Brain Research* **40**, 385–393.

Van Woert, M.H., Rosenbaum, D., Howieson, J. and Bowers, M.B. (1977). *New Engl. J. Med.* **296**, 70–75.

Vial, H., Claustre, Y. and Pacheco, H. (1974). *J. Pharmacol. (Paris)* **5**, 461–478.

Wada, J.A., Balzamo, E., Meldrum, B.S. and Naquet, R. (1972). *Electroenceph. clin. Neurophysiol.* **33**, 520–526.

Waksman, A., Rubinstein, M.K., Kuriyama, K. and Roberts, E. (1968). *J. Neurochem.* **15**, 351–357.

Williams, D. and Russell, W.R. (1941). *Lancet* **1**, 476–479.

Winters, W.D. and Spooner, E.C. (1965). *Int. J. Neuropharmacol.* **4**, 197–200.

Woodbury, D.M. (1969). *In* "Basic Mechanisms of the Epilepsies" (eds. Jasper, H.H., Ward, A.A. and Shute, A.), pp. 647–688. Churchill, London.

Wu, J-Y., and Roberts, E. (1974). *J. Neurochem.* **23**, 759–767.

Young, A.B., Zukin, S.R. and Snyder, S.H. (1974). *Proc. Nat. Acad. Sci.* **71**, 2246–2250.

Zukin, S.R., Young, A.B. and Snyder, S.H. (1974). *Proc. Nat. Acad. Sci.* **71**, 4802–4807.

NEUROTRANSMITTERS AND SENILE DEMENTIA

D.M. BOWEN

Department of Neurochemistry, Institute of Neurology,
Queen Square, London, U.K.

Introduction

The aetiology of senile and presenile dementia is largely unknown,
and at present treatment can be offered only to that small proportion
of cases in which some underlying disease is identified. However, the
history of myasthenia gravis and Parkinson's disease shows that
effective treatment may precede an understanding of aetiology by
many years, and in both these conditions such treatment has been
directed towards a defect in synaptic transmission. We have therefore
investigated the brains of patients with dementia to see if there was
evidence for selective fall-out of any one type of cortical neurone: if
there were, symptomatic treatment might be possible.

The diseased brains we have studied were from patients with a
mean age of 86 (range 79–97) and a provisional clinical diagnosis of
senile dementia; the morphological cerebral changes were consistent
with a definitive diagnosis of senile dementia (Diagnostic Group 0: 2
of Corsellis (1962); i.e. the brains showed marked 'senile'
morphological changes in the cerebral cortex, including the
hippocampus, but little or no evidence of cerebrovascular disease).
There is no discernible pathological difference between Alzheimer's
disease and senile dementia.

Cell Loss

Since assessment of loss of nerve cells by histological methods presents
many problems, biochemical methods have been applied to the
analysis of whole temporal lobe. The diseased material has been
compared to matched controls. Gross atrophy (18% reduction in
brain weight) of the lobe was accompanied by larger decreases (about
33%) in enzyme, lipid, nucleic acid and protein markers of nerve cells
(Bowen and Davison, 1976). Other estimations suggest that deposition
of amyloid (and/or mucopolysaccharides) and glial reactions (e.g.
microgliosis) may account for the selective preservation of brain

weight. Such changes are not seen in the temporal lobe of non-demented subjects. Thus in both non-vascular and in the less prevalent multi-infarct dementias the biochemical evidence is consistent with a loss of about half the concentration of neuronal 'markers' (Table I). This agrees with the morphological observations

TABLE I

Changes in Biochemical Indices of Brain Degeneration in Temporal Lobes from Cases of Senile Dementia

Parameter	Measured as potential in index of	% Reduction from elderly control value	P value
Total protein	—	20.7	< 0.01
β-galactosidase	Neuronal perikarya	20.9	< 0.1
β-glucuronidase	Glial proliferation and hypertrophy	0.0	NS
Ganglioside NANA	Nerve cell membranes	28.1	< 0.001
Choline acetyltransferase	Cholinergic neurones	58.2	< 0.001
Acetylcholinesterase	Cholinergic nerve endings and post-synaptic membranes	46.9	< 0.1
Atropine binding	Muscarinic-cholinergic post synaptic receptor density	0.0	NS
Glutamate decarboxylase	GABA-ergic neurones	65.3	< 0.05
Aromatic amino acid decarboxylase	Monoaminergic neurones	59.8	< 0.1
2' 3'-Cyclic nucleotide 3-phosphohydrolase	Myelinated axons	34.9	< 0.001

of Colon (1973) who assessed loss of neurons in the neocortex as about 57%. If there are such losses, are they selective and is the synapse particularly affected?

Selective Vulnerability

In certain other abiotrophies there is evidence that specific types of neurones are selectively affected in discrete parts of the nervous system (e.g. dopaminergic neurones in the substantia nigra in idiopathic Parkinson's disease and gabanergic neurones in the striatum in Huntington's chorea). Despite these important advances the investigation of 'markers' of specific types of neurones or synapses is still very much in its infancy. Furthermore, although cholinergic

(Waser, 1975) and gabanergic (Roberts *et al*, 1976) nerve cells are
well recognised, there are indications that peptides might also be
involved in synaptic functioning, particularly in the cerebral cortex.
The 'marker' techniques that are being currently investigated include
the measurement of the concentration of neurotransmitters in post-
mortem brains. Although this approach is useful for dopamine it is
not applicable to acetylcholine and may be invalid for GABA
(Bachelard *et al*, 1976). Results obtained using CSF are confounded
by several epiphenomena, including brain atrophy, which can cause
an increase in ventricular volume and presumably a corresponding
decrease in neurotransmitter concentration. Other techniques include
the assay and immunochemical localisation of neurotransmitter-
related and 'second messenger' enzymes (i.e. the nucleotide cyclases);
quantifying neurotransmitter-receptor proteins; and measuring *in
vitro* high affinity uptake of neurotransmitters. Due to terminal or
agonal and post-mortem effects it appears that many of these
techniques can only be meaningfully applied to biopsy specimens
and experimental animal models. Since there are no reliable animal
models of either presenile or senile dementia the investigation of
cortical biopsies is at present the only realistic approach that may
directly solve the problem of selective vulnerability in these
dementias.

Hypoxia

It is unclear whether or not our earlier finding of reduced cortical
GAD activity in Alzheimer's disease (group 5, Table II) accurately
reflects the pre-terminal state of GABA-containing neurones (Bowen
et al, 1976). Patients with Alzheimer's disease often have a defective
respiratory reflex (Allison, 1962) with reduced cerebral blood flow
terminally (Ingvar and Gustafson, 1970). This is important, for GAD
activity and the concentration of a soluble acidic brain protein
(neuronin S-6) seem to decline both in baboon brain with reduced
cerebral blood flow (Bowen *et al*, 1976) and in human brains from
non-demented patients either dying of bronchopneumonia (compare
GAD activity in groups 3 and 4, Table II) or with evidence of more
severe terminal cerebral hypoxia (Bowen *et al*, 1976). Furthermore,
the concentration of neuronin S-6 is markedly depleted in cortex from
cases of Alzheimer's disease (Bowen *et al*, 1976; Smith and Bowen,
1976). There are indications (Bowen *et al*, in press) that decreases in
GAD activity may be due to denaturation at acidic pH values. Our
findings, and those of McGeer and McGeer (1976) on the effects of
terminal coma, demonstrate the importance of making allowance for

TABLE II

Choline Acetyltransferase (CAT) and Glutamate Decarboxylase (GAD) Activities in Human Post-operation and Post-mortem Neocortical Specimens

Brain specimen and group no.	CAT activity (units per min)	GAD activity (units per min)	Total protein (mg per g wet weight)
Post-operation specimens			
1) Control tissue*	66.80 ± 26.96 (10)	1333 ± 467 (8)	81.2 ± 17.5 (10)
2) Alzheimer's disease and related disorders†	27.52 ± 15.38 (6)[a]	1393 ± 329 (5)	75.7 ± 17.6 (6)
*Post-mortem tissue***			
3) Controls dying suddenly	62.78 ± 14.75 (12)	1090 ± 460 (17)	74.6 ± 14.9 (17)
4) Controls dying of bronchopneumonia	71.83 ± 16.42 (7)	690 ± 182 (8)[c]	75.1 ± 18.4 (7)
5) Alzheimer's disease dying of bronchopneumonia	36.62 ± 10.65 (6)[b]	295 ± 89 (9)[d]	75.1 ± 10.4 (8)

CAT, GAD and protein were measured as previously described (Bowen et al, 1976). Data are mean ± s.d.; numbers in brackets are number of cases; enzyme units are pmol acetylcholine or GABA formed per mg protein. [a]–[d] identify significant differences between closely matched groups. CAT activity: [a]group 2 compared with 1, $P < 0.01$; [b]group 5 compared with 4, $P < 0.001$. GAD activity: [c]group 4 compared with 3, $P < 0.05$; [d]group 5 compared with 4, $P < 0.001$.

*Normal neocortex removed at surgery for tumours; mean age 42 ± 20 range 10–67). For these specimens neither CAT nor GAD are age-dependent which contrasts with data for post-mortem samples from cases aged 5–50 (McGeer and McGeer, 1976).

†Neocortex from 2 biopsy confirmed cases of Alzheimer's disease and 2 probable cases; an adolescent with cerebral atrophy and some symptoms of schizophrenia ('dementia praecox') and a case of Parkinson's disease with a sister with pre-senile dementia; mean age 49 ± 18 (range 18–65).

**Calculated from previously published data for neocortex from cases matched with respect to the interval between death and autopsy. The ages for groups 3, 4 and 5 are 72 ± 13 (range 50–100), 66 ± 16 (range 37–84) and 85 ± 5 (range 79–97). For these specimens neither CAT nor GAD activities are age-dependent (Bowen et al, 1976).

the terminal state.

Cholinergic Neurones

In an attempt to resolve these difficulties we have examined cortical biopsies removed at craniotomy. These studies suggest that CAT but not GAD activity is often reduced in cortex from 'demented' patients with cerebral atrophy (group 2, Table II). Of these cases, four were Alzheimer's disease. In these four alone, CAT activity (24.10 ± 18.63 units) is reduced by a mean of 64% (P < 0.02) while GAD activity (1298 ± 205 units) is unchanged. The results for the biopsy samples are unique since it is the first time that CAT activity has been shown to be reduced in diseased human brain when the activity of GAD is typical of the mean control value. Comparable results have been reported recently (Enna *et al*, 1976) for the density of putative acetylcholine and GABA-receptor binding sites in the striatum in Huntington's chorea.

The depletion in CAT activity that we have detected is of particular interest because the anticholinergic glycolate esters, of which the muscarinic cholinergic marker quinuclidinyl benzilate (Enna *et al*, 1976) is a prominent example, can cause complete amnesia (Abood, 1968). Furthermore, despite inconsistencies in the literature (Karczmar, 1975), at least one pharmacological study (Drachman and Leavitt, 1974) on humans indicates that the cholinergic system is involved in age-related memory degeneration. In order to provide a rational basis of treatment, other tentative markers of the state and number of specific types of neurones are being measured. Preliminary results for another index of the integrity of the cholinergic system, high affinity choline uptake, have shown that the reduced uptake in cortex from a 63 year old patient who probably had Alzheimer's disease was similar to the reduction of CAT activity (66% and 67% respectively). In contrast, GAD activity and β-alanine independent GABA uptake (Bowen *et al*, 1976) together with two potential indices of post synaptic membranes, guanylate and adenylate cyclase activities (Bowen *et al*, in press) were all within 1.5 s.d. of the appropriate mean control value.

Conclusions

When the data for the biopsy and autopsy samples are considered together they suggest that at least the presynaptic cholinergic system is selectively affected in Alzheimer's disease. This perhaps offers the possibility of treatment with a centrally acting anticholinesterase, for at present only about 14% of cases of organic dementia are treatable (Pearce and Miller, 1973). This therapeutic approach may not be entirely successful, however, as there are indications that other aspects of brain metabolism may be decreased, which may interact with the presynaptic cholinergic system (Bowen and Davison, in press).

References

Abood, L.G. (1968). *In* "Drugs Affecting the Central Nervous System" (ed. Burger, A.), pp. 127–167. Edward Arnold, London.

Allison, R.S. (1962). The Senile Brain. Edward Arnold, London.

Bowen, D.M. and Davison, A.N. (1976). *In* "Biochemistry and Neurological Disease" (ed. Davison, A.N.). Blackwells.

Bowen, D.M., Smith, C.B., White, P. and Davison, A.N. (1976). *Brain* **99**, 459–496.

Bowen, D.M. *et al.* (1976). *Brain Res.* **117**, 503–507.

Bowen, D.M. *et al. Brain* (in press).

Bowen, D.M. and Davison, A.N. *In* "Recent Advances in Geriatric Medicine" (ed. Isaacs, B.). Churchill Livingstone, Edinburgh (in press).

Colon, E.J. (1973). *Acta neuropath. (Berl.)* **23**, 281–290.

Corsellis, J.A.N. (1962). Mental Illness and the Ageing Brain. O.U.P.

Drachman, D.A. and Leavitt, J. (1974). *Arch. Neurol.* **30**, 113–121.

Enna, S.J. *et al.* (1976). *New Engl. J. Med.* **294**, 1305–1309.

Ingvar, D.H. and Gustafson, L. (1970). *Acta neurol. scand. Suppl. 43* **46**, 42–63.

Karczmar, A.G. (1975). *In* "Cholinergic Mechanisms (ed. Waser, P.G.), pp. 506–508. Raven, New York.

McGeer, P.L. and McGeer, E.G. (1976). *J. Neurochem.* **26**, 65–76.

Pearce, J. and Miller, E. (1973). *In* "Clinical Aspects of Dementia". Bailliere Tindall, London.

Roberts, E., Chase, T.N. and Tower, D.B. (eds.). (1976). GABA in Nervous System Function. Raven Press, New York.

Smith, C.B. and Bowen, D.M. (1976). *J. Neurochem.* **27**, 1521–1528.

Waser, P.G. (ed.). (1975). Cholinergic Mechanisms. Raven Press, New York.

Weintraub, S.T., Modak, A.T. and Stavinoha, W.B. (1976). *Brain Res.* **105**, 179.

THE BIOCHEMICAL AND BEHAVIOURAL EFFECTS OF GIVING A MONOAMINE OXIDASE INHIBITOR AND L-TRYPTOPHAN TO RATS

C.A. MARSDEN

Department of Physiology and Pharmacology, The Medical School, Clifton Boulevard, Nottingham, U.K.

Introduction

There is convincing evidence that 5-hydroxytryptamine (5HT) is a neurotransmitter in nerve fibres which are derived from cell bodies in the midbrain raphe nuclei, and which terminate in various brain regions, such as the striatum, hypothalamus, and cortex, and also in the spinal cord (see Fuxe and Jonsson, 1974). Functionally 5HT has been implicated in the control of sexual behaviour, in the sleep-waking cycle and in reactivity responses, and it may possibly be involved in various clinical conditions including affective disorders and schizophrenia (e.g. see Green and Grahame-Smith, 1975, also Shaw, this volume, p. 201, and Crow and Johnstone, this volume, p. 207). This makes it important to understand the mechanisms controlling post-synaptic 5HT receptor activity and to identify specific drugs that influence this activity.

One way to look at receptor activity is to use a behavioural response which can be elicited by receptor stimulation. A model that has been used extensively in recent years is the behavioural response produced by giving L-tryptophan, a precursor of 5HT (Fig. 1), to rats which have been pretreated with an inhibitor of monoamine oxidase (MAO) (Grahame-Smith, 1971a). This produces a behavioural syndrome consisting of hyperactivity, hyperreactivity, reciprocal forepaw treading, lateral head weaving, straub tail, hind limb abduction, tremor and hyperpyrexia. This combination of MAO inhibitor and L-tryptophan has been used not only as a behavioural model of 5HT receptor activity in animals (see Green and Grahame-Smith, 1976a; Jacobs, 1976) but also successfully in the treatment of depression in man (e.g. Coppen, Shaw and Farrell, 1963).

In the first part of this chapter we will review the evidence that the behavioural syndrome is the result of 5HT receptor activity. We will then discuss recent evidence suggesting that 5HT may not be the only metabolite of tryptophan involved in the syndrome. Lastly the possible clinical implications of this experimental model will be considered.

METABOLISM OF 5-HT & TRYPTAMINE

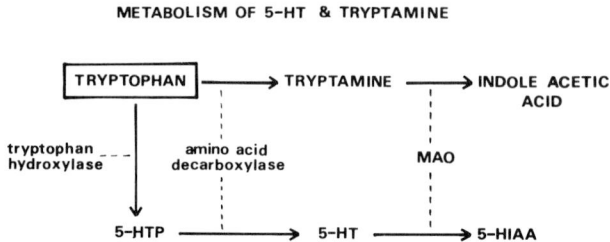

Fig. 1   Metabolism of tryptophan to 5HT and tryptamine.

Evidence that the Syndrome Involves 5HT Receptor Stimulation

5HT is formed by the hydroxylation of tryptophan to 5-hydroxy-tryptophan (5-HTP), and the subsequent decarboxylation of 5-HTP (Fig. 1). Its major route of catabolism is oxidation by MAO to 5-hydroxyindoleacetic acid (5HIAA). Tryptophan hydroxylase is unsaturated at normal brain tryptophan concentrations, so procedures that increase the concentration of brain tryptophan consequently increase brain 5HT turnover. When L-tryptophan is given to rats there is a marked increase in brain 5HIAA and a lesser increase in 5HT, but none of the behavioural changes mentioned above are seen. However if the rats are pretreated with an MAO inhibitor, preventing the catabolism of 5HT to 5HIAA, and are then given L-tryptophan, there is a marked accumulation of 5HT; and it is under these conditions that the behavioural syndrome is apparent. Grahame-Smith (1971a) has suggested that the behaviour only occurs because 5HT neuronal stores are full and the accumulated 5HT therefore 'spills over' onto the post-synaptic receptors; this does not happen when L-tryptophan is given alone, because excess 5HT synthesis is rapidly corrected by MAO activity.

The evidence that the syndrome is mediated by stimulation of 5HT receptors is mainly pharmacological, and is summarised in Table I. The syndrome is produced by pharmacological manipulations that enhance the amount of 5HT available to receptors. Thus the administration of tryptophan, 5-hydroxytryptophan (5HTP) or a 5HT neuronal re-uptake blocker to rats pretreated with an MAO inhibitor can all cause the syndrome (Grahame-Smith, 1971a; Modigh, 1972; Holman *et al*, 1976) but it does not occur if the rats are previously given an inhibitor of either tryptophan hydroxylase or central 5HTP

TABLE I

MAO Inhibitor + Tryptophan Induced Behaviour

Behavioural Syndrome [1,2]	Produced by	Prevented by
Hyperactivity	MAOI + TP[1,2]	Inhibitor of 5HT synthesis, e.g. p-chlorophenylalanine, [1] NSD 1055. [1]
Hyper-reactivity	MAOI + 5HTP[3]	5HT receptor blockers, e.g. methergoline[9]
Lateral head weaving	MAOI + 5HT re-uptake blocker[4]	(β adrenergic blockers)[10]
Forepaw treading	5-Methoxy-N, N-dimethyl-tryptamine[5]	Inhibitors of DA synthesis, e.g. α-methyl-p-tyrosine[11]
Straub tail	Fenfluramine[6]	DA receptor blockers, e.g. haloperidol, [12] chlorpromazine, [12] α-flupenthixol[12].
Tremor	p-chloramphetamine[6]	
Hyperpyrexia	MAOI + tryptamine[7]	
	α-methyltryptamine[8]	

1. Grahame-Smith, 1971a.
2. Jacobs, 1974.
3. Modigh, 1972.
4. Holman *et al*, 1976.

5. Grahame-Smith, 1971b.
6. Trulson and Jacobs, 1976.
7. Foldes and Costa, 1975.
8. Marsden and Curzon, 1977.

9. Crow and Deakin, 1977.
10. Green and Grahame-Smith, 1976.
11. Green and Grahame-Smith, 1974.
12. Heal *et al*, 1976.

MAOI = Monoamine oxidase inhibitor.
TP = L-tryptophan.

decarboxylase (Grahame-Smith, 1971a), so preventing 5HT synthesis. The direct agonist of 5HT receptors, 5-methoxy-N, N-dimethyl-tryptamine (Grahame-Smith, 1971b), and two drugs believed to cause release of 5HT, p-chloroamphetamine and fenfluramine (Trulson and Jacobs, 1976), also produce the syndrome. Pretreatment with an inhibitor of tryptophan hydroxylase (e.g. p-chlorophenylalanine) to deplete brain 5HT prevents p-chloroamphetamine from eliciting the syndrome. Interestingly p-chloroamphetamine initially causes a release of neuronal 5HT and then produces a period of prolonged 5HT depletion. The initial period is associated with the behavioural syndrome, while during the later period the behavioural effects are similar to those produced by other drugs that deplete 5HT (Sheard and Davis, 1976; Zemlan *et al*, 1977). Methergoline, a specific antagonist at 5HT receptors, blocks most of the syndrome produced by either an MAO inhibitor plus L-tryptophan or by p-chloroam-phetamine, but does not block the hyperactivity or hyperreactivity (Crow and Deakin, 1977). Recently Green and Grahame-Smith (1976b) have shown that L-propranolol and other β-adrenergic blocking drugs prevent the hyperactivity in rats given an MAO inhibitor and L-tryptophan, but do not alter the hyperactivity produced by an MAO inhibitor and L-dopa. This suggests that β-adrenergic blockers also inhibit 5HT neurotransmission. Jacobs and Klemfuss (1975) have partially identified the 5HT neuronal systems that mediate the syndrome by making a series of complete brain transections and showing that the tremor, forepaw treading, straub tail and lateral head weaving are mediated by neural mechanisms present in the pons, medulla and spinal cord. This finding highlights the importance of the 5HT cell bodies in the medullary raphe nuclei and their synaptic contacts in the spinal cord (Jacobs, 1976).

Involvement of Dopaminergic Neurones

It is also apparent that the syndrome is influenced by dopaminergic neuronal activity, as both the inhibitor of tyrosine hydroxylase, α-methyl-p-tyrosine, and most dopaminergic blocking drugs prevent the hyperactivity (Green and Grahame-Smith, 1974; Heal *et al*, 1976). L-dopa given to rats pretreated with an MAO inhibitor produces many of the features seen after an MAO inhibitor plus L-tryptophan, together with more obvious 'catecholamine signs' (Jacobs, 1976). p-Chloro-phenylalanine, an inhibitor of 5HT synthesis, blocks the effects of L-dopa, indicating that the latter may act through 5HT neuronal systems by causing the release of endogenous 5HT (Jacobs, 1974). Green and Kelly (1976), however, found that although

p-chlorophenylalanine did block the MAO inhibitor plus L-dopa
response this was because p-chlorophenylalanine inhibited the uptake
of L-dopa by brain rather than because of its effect on 5HT synthesis.

There is thus strong evidence that, with the possible exception of
the hyperactivity and reactivity, the behavioural syndrome produced
by L-tryptophan administration to MAO inhibitor pretreated rats is
caused by the stimulation of 5HT receptors located principally in the
lower brain stem and spinal cord. Although the response can be
modulated by dopaminergic neurones the mechanism by which this
occurs is not understood.

Possible Involvement of Tryptamine in the Syndrome

Although increased accumulation of 5HT seems an essential feature
of this syndrome, it is possible that another metabolite of tryptophan
or 5HT could be involved, acting either directly on 5HT receptors or
indirectly by increasing the amount of 5HT available to the receptors.
The syndrome could depend upon the synthesis of an N-substituted
derivative of 5HT (Squires, 1975). There is however no direct evidence
for the synthesis of such derivatives when tryptophan is given to
MAO inhibitor pretreated rats. More is known about factors influencing
the synthesis of tryptamine, which is formed by the direct decarboxylation of tryptophan and is very rapidly metabolised by MAO (Fig. 1).

Endogenous tryptamine is found in rat and human brains (Saavedra
and Axelrod, 1973). In the rat, brain tryptamine levels are markedly
increased when MAO is inhibited (Marsden and Curzon, 1974;
Tabakoff *et al*, 1977) but the most striking increase is when tryptophan
is given to MAO inhibitor pretreated rats, when there is a proportionately greater increase in tryptamine than in 5HT (Marsden and Curzon,
1974; Tabakoff *et al*, 1977). The increase on giving L-tryptophan
alone is very small (Saavedra and Axelrod, 1973; Marsden and
Curzon, 1974). Furthermore giving tryptamine ($1-5$ mg/kg) to rats
pretreated with an MAO inhibitor produces hyperactivity similar to
that seen on giving L-tryptophan to similarly pretreated rats (Foldes
and Costa, 1975).

We have recently looked at the possible role of tryptamine in the
behavioural response to an MAO inhibitor plus L-tryptophan, using
both motor activity measurements and a behavioural score (Fig. 2)
to determine the intensity of the syndrome (Curzon and Marsden,
1977; Marsden and Curzon, 1978).

In the first experiment, we administered a dose of tryptamine (0.75
mg/kg), which by itself does not produce the syndrome, at the same
time as L-tryptophan (50 mg/kg), to rats previously given the MAO

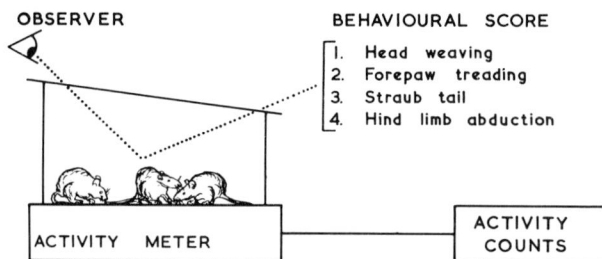

Fig. 2 Method used to measure behaviour produced by tranylcypromine (20 mg/kg) plus tryptophan in rats. Tryptophan was administered 30 minutes after tranylcypromine, and activity was counted and behaviour scored for 60 minutes. Rats were then killed, and brain tryptophan, 5HT and tryptamine determined (Marsden and Curzon, 1978).

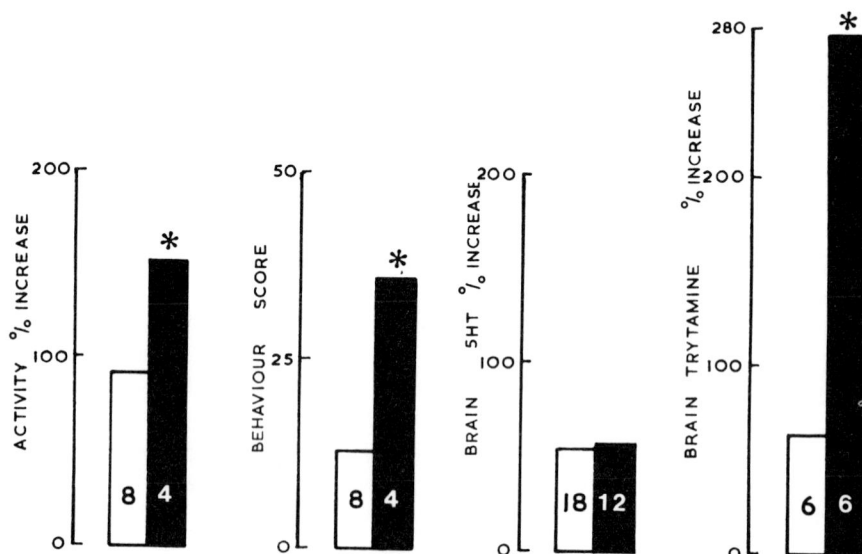

Fig. 3 Effects of L-tryptophan 50 mg/kg (open columns) and L-tryptophan 50 mg/kg + tryptamine 0.75 mg/kg (black columns) on activity, behaviour score, brain 5HT and tryptamine in rats pretreated with tranylcypromine (20 mg/kg). L-Tryptophan or L-tryptophan plus tryptamine were given 30 minutes after tranylcypromine and rats killed 60 minutes later. Behaviour was scored every 15 minutes during the 60 minutes after L-tryptophan (+/− tryptamine). Results are given as per cent increase over t tranylcypromine alone except for the behavioural score, which is in absolute terms as rats given tranylcypromine show none of the components of the behavioural score.
Number of determinations indicated in the columns.
*P < 0.001.

inhibitor tranylcypromine (20 mg/kg). We found that it enhanced the response seen with tryptophan; it did not alter the increase in brain 5HT but it did cause a large increase in brain tryptamine (Fig. 3). We then compared the behavioural effects of two doses of L-tryptophan (50 and 100 mg/kg) given to rats pretreated with tranylcypromine, and found that the larger dose produced a small increase in motor activity but a greater increase in the behavioural score. This was associated with the proportionately greater increase in brain tryptamine than in 5HT (Fig. 4). Both these results suggest that increased brain tryptamine may be a factor in the behavioural syndrome.

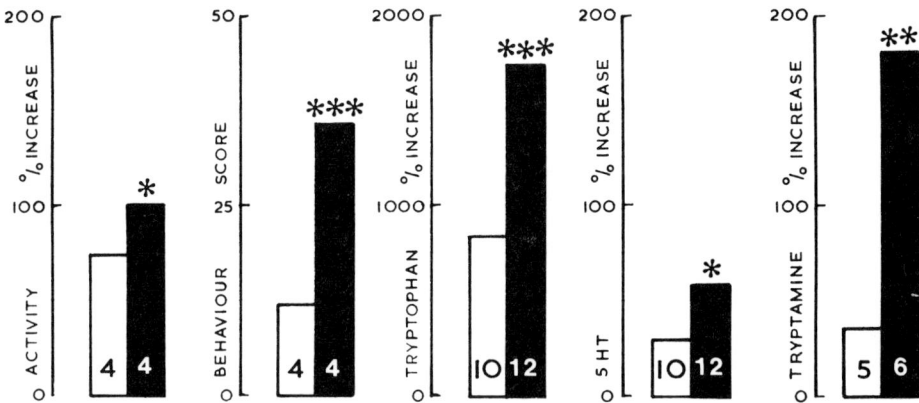

Fig. 4 Effect of L-tryptophan 50 mg/kg (open columns) and 100 mg/kg (black columns) on activity, behaviour score, brain tryptophan, 5HT and tryptamine in rats pretreated with tranylcypromine 20 mg/kg. L-Tryptophan was given 30 minutes after tranylcypromine and rats were killed 60 minutes later. Behaviour was scored every 15 minutes during the 60 minutes after tryptophan administration. Results are given as per cent increase over tranylcypromine alone except for the behaviour score, which is the absolute score. Number of determinations indicated in the columns.
* $P < 0.05$; ** $P < 0.01$; *** $P < 0.001$.

Tryptamine can be synthesised both inside and outside the brain and, unlike 5HT, is able to cross the blood/brain barrier (Green and Sawyer, 1960). This makes it theoretically possible for tryptamine formed outside the brain to influence behaviour via a central mechanism. To test this we gave rats a peripheral decarboxylase inhibitor (R04-4602, 50 mg/kg) 30 minutes before tranylcypromine, which was followed in the usual way after 30 minutes by L-tryptophan (100 mg/kg). The peripheral decarboxylase inhibitor prevents the synthesis of tryptamine outside the brain without affecting its synthesis in the brain. Under these conditions the L-tryptophan produced less marked hyperactivity and a lower behavioural score; there was no change in brain 5HT but there was a marked fall in brain tryptamine (Fig. 5). Thus it appears that tryptamine formed peripherally can affect behaviour.

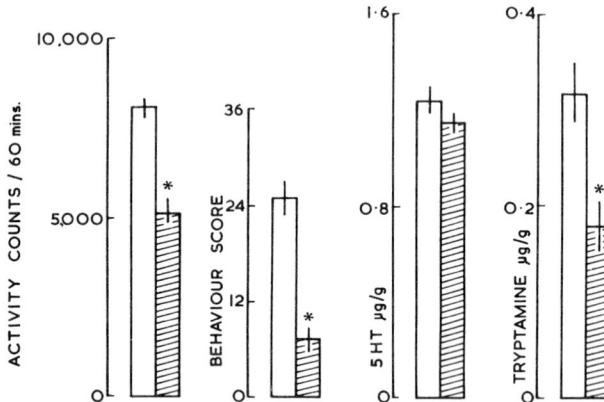

Fig. 5 Effect of inhibiting peripheral decarboxylase on activity, behaviour score and brain 5HT and tryptamine of rats given tranylcypromine (20 mg/kg) 30 minutes before L-tryptophan (100 mg/kg). The rats were killed 60 minutes after tryptophan administration. R04-4602 (50 mg/kg) was given 30 minutes before tranylcypromine. Behaviour was scored every 15 minutes during the 60 minutes after L-tryptophan administration.
Open columns: tranylcypromine + L-tryptophan.
Hatched columns: R04-4602, tranylcypromine + L-tryptophan.
Number of determinations: activity and behaviour: 4; 5HT: 12; tryptamine: 6.
*P < 0.01 v tranylcypromine + L-tryptophan.

The next question is whether tryptamine influences the behavioural response caused by an MAO inhibitor plus L-tryptophan by acting directly on post-synaptic 5HT receptors, or indirectly by increasing the concentration of 5HT at these sites. When brain 5HT is depleted by inhibiting tryptophan hydroxylase the behavioural response to tryptamine is reduced (Foldes and Costa, 1976). Now this might indicate that tryptamine acts through the release of 5HT, but it is also consistent with the idea that tryptamine acts directly on 5HT receptors but is less available to them because it is being taken up in increased amounts by neurones depleted of 5HT. Tryptamine is a competitive inhibitor of 5HT uptake (Horn, 1973).

Brain tryptamine is synthesised not only in 5HT neurones but also in dopamine neurones (Marsden and Curzon, 1974), because tryptophan is decarboxylated by the same enzyme that decarboxylates 5-hydroxytryptophan and dopa. Furthermore the uptake of tryptophan is not specific to 5HT neurones. Therefore the effects of tryptamine on behaviour may not necessarily be mediated solely by

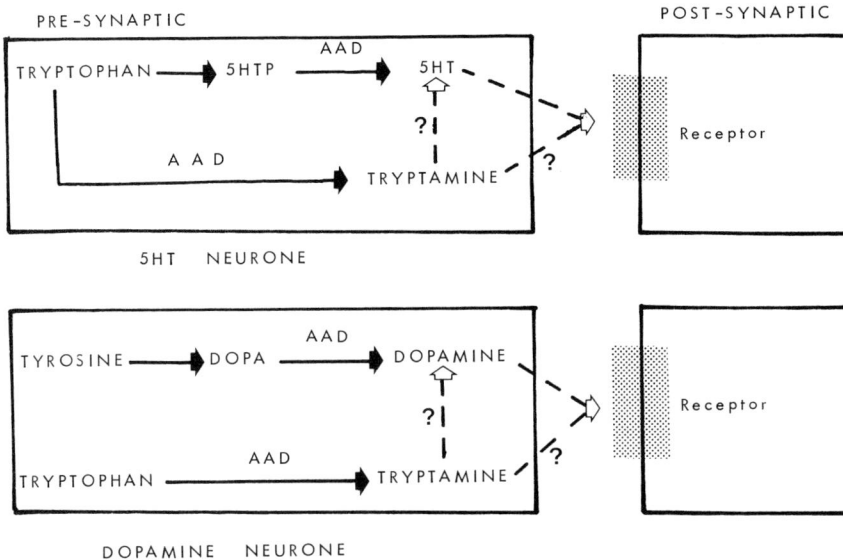

Fig. 6   Diagram showing how tryptamine could be synthesised in both 5HT and catechol-aminergic neurones, due to the ubiquitous distribution of amino acid decarboxylase and the non-specificity of the tryptophan uptake process. Tryptamine may displace stored endogenous 5HT or dopamine by competing for storage sites.

The solid lined arrows show synthetic pathways and the broken lined arrows show the release, possible release, or possible modulation of release, of 5HT, tryptamine and dopamine. AAD = Amino acid decarboxylase.

its effects on 5HT neuronal systems but may also involve dopaminergic neurones, through a release of dopamine. 6-Hydroxydopamine, which causes degeneration of dopamine neurones, reduces the effects of tryptamine on behaviour (Foldes and Costa, 1975). This might explain why the 5HT receptor antagonist methergoline blocked the syndrome produced by an MAO inhibitor plus L-tryptophan apart from the hyperactivity and reactivity (Crow and Deakin, 1977). The components unaffected by 5HT receptor blockade may be caused by the release of dopamine on to dopamine receptors following an increase in tryptamine within dopaminergic neurones (Fig. 6).

In summary, there is good evidence that the behavioural syndrome produced by an MAO inhibitor plus L-tryptophan is dependent upon the presence of adequate amounts of 5HT, and upon the stimulation of 5HT receptors, and that catecholaminergic neurones are also involved. There is also an increase in the synthesis of tryptamine, which may influence the behavioural response either by direct stimulation of 5HT receptors or indirectly via the release of 5HT and possibly dopamine. This raises the question of whether tryptamine plays a role under physiological conditions, and whether it might modulate the release of 5HT at synapses (Boulton, 1976); also whether tryptamine dysfunction is a factor in psychiatric illness.

Clinical Implications

The effectiveness of tryptophan in the treatment of affective disorders is controversial (Coppen *et al*, 1972; Jensen *et al*, 1975; Murphy *et al*, 1974; Dunner and Fieve, 1975) but the chronic administration of MAO inhibitors in conjunction with L-tryptophan has proved effective (e.g. Coppen *et al*, 1963). The MAO inhibitors are given at much lower doses than in the rat experiments, but the tryptophan dose is comparable. Furthermore, the administration of tryptophan together with imipramine is reported to be better than imipramine alone (Walinder *et al*, 1975). Both these treatments will increase the amount of 5HT available to receptors, and it seems reasonable to believe that the development of specific 5HT releasers and selective 5HT agonists (Green and Grahame-Smith, 1976a) should be rewarding in the treatment of depression, especially if it becomes clinically possible to isolate a subgroup of depressives with disturbed 5HT metabolism (Asberg *et al*, 1976). It will also be interesting to know, under the clinical conditions in which these drug combinations are used, the relative effects on brain 5HT and tryptamine metabolism, as reflected by changes in their metabolites in either lumbar or ventricular CSF.

References

Asberg, M., Thonen, P., Traskman, L., Bertilsson, L. and Ringberger, V. (1976). *Science* **191**, 478–480.

Boulton, A.A. (1976). *In* "Trace Amines and the Brain" (eds. Usdin, E. and Sandler, M.), pp. 21–39. M. Dekker, New York.

Coppen, A., Shaw, D.M. and Farrell, J.P. (1963). *Lancet* **1**, 79–81.

Coppen, A., Whybrow, P.C., Noguera, R., Maggs, R. and Prange, A.J. Jr. (1973). *Arch. gen. Psychiat.* **26**, 234–241.

Crow, T.J. and Deakin, J.F.W. (1977). *Brit. J. Pharmac.* **59**, 461p.

Curzon, G. and Marsden, C.A. (1977). *Brit. J. Pharmac.*, 307–308p.

Dunner, D.L. and Fieve, R.R. (1975). *Am. J. Psychiatry* **132**, 180–183.

Foldes, A. and Costa, E. (1975). *Biochem. Pharmac.* **24**, 1617–1621.

Fuxe, K. and Jonsson, G. (1974). *Adv. Biochem. Pharmacol.* **10**, 1–12.

Grahame-Smith, D.G. (1971a). *J. Neurochem.* **18**, 1053–1066.

Grahame-Smith, D.G. (1971b). *Brit. J. Pharmac.* **43**, 856–864.

Green, A.R. and Grahame-Smith, D.G. (1974). *Neuropharmacology* **13**, 949–959.

Green, A.R. and Grahame-Smith, D.G. (1975). Handbook of Psychopharmacology 3, 169–245. Plenum, New York.

Green, A.R. and Grahame-Smith, D.G. (1976a). *Nature* **260**, 487–491.

Green, A.R. and Grahame-Smith, D.G. (1976b). *Nature* **262**, 594–596.

Green, A.R. and Kelly, P.H. (1976). *Brit. J. Pharmac.* **57**, 141–147.

Green, H. and Sawyer, J.L. (1960). *Proc. Soc. Exp. Biol. Med.* **104**, 153–155.

Heal, D.J., Green, A.R., Boullin, D.J. and Grahame-Smith, D.G. (1976). *Psychopharmacology* **49**, 287–300.

Holman, B., Seagraves, E., Elliot, G.R. and Barchas, J.D. (1976). *Behavioural Biology* **16**, 507–514.

Horn, A.S. (1973). *J. Neurochem.* **21**, 883–888.

Jacobs, B.L. (1974). *Psychopharmacologia* **39**, 81–86.

Jacobs, B.L. (1976). *Life Sciences* **19**, 777–786.

Jacobs, B.L. and Klemfuss, H. (1975). *Brain Res.* **100**, 450–457.

Jensen, K., Fruensgaard, K., Ahlfors, U-G., Pihkanen, T.A., Tuomikoski, S., Ose, E., Dencker, S.J., Lindberg, D. and Nagy, A. (1975). *Lancet* **11**, 920.

Marsden, C.A. and Curzon, G. (1974). *J. Neurochem.* **23**, 1171–1176.

Marsden, C.A. and Curzon, G. (1977). *Neuropharmacology* **16**, 489–494.

Marsden, C.A. and Curzon, G. (1978). *Psychopharmacology* (in press).

Modigh, K. (1972). *Psychopharmacologia (Berl.)* **23**, 48–54.

Murphy, D.L., Baker, M., Goodwin, F.K., Miller, H., Kotin, J. and Bunney, W.E. Jr. (1974). *Psychopharmacologia (Berl.)* **34**, 11–20.

Saavedra, J.M. and Axelrod, J. (1973). *J. Pharmac. exp. Ther.* **185**, 523–529.

Sheard, M.H. and Davis, M. (1976). *Eur. J. Pharmacol.* **40**, 295–302.

Squires, R.F. (1975). *J. Neurochem.* **24**, 47–50.

Tabakoff, B., Moses, F., Philips, S.R. and Boulton, A.A. (1977). *Experientia* **33**, 380–381.

Trulson, M.E. and Jacobs, B.L. (1976). *Eur. J. Pharmacol.* **36**, 149–154.

Zemlan, F.P., Trulson, M.E., Howell, R. and Hoebel, B.G. (1977). *Brain Res.* **123**, 347–356.

5-HYDROXYTRYPTAMINE AND DEPRESSION

D.A. SHAW

Biochemical Psychiatry Laboratory, Department of Psychological Medicine, Whitchurch Hospital, Cardiff, U.K.

Introduction

The affective disorders are phasic illnesses in which periods of normality alternate with episodes of abnormality. Perhaps it would be more accurate to think of them as conditions in which there is a change in vitality rather than to associate them in a more limited way with depressive or manic labels.

In the phase of lowered vitality, any one patient may show several of the following symptoms: depression, often with diurnal variation of mood; loss of drive for work, food and sex; anxiety; diminished energy, interest and ability to concentrate; disturbed sleep, often with early morning wakening; and secondary psychological symptoms such as suicidal thoughts, hypochondriasis, paranoid ideas, self-blame, delusions of catastrophe, etc.

Biochemical Basis for Depression

Most authorities now accept that the affective disorders are biochemically determined disturbances of brain function, and this is a view which has come from a variety of observations.

First, susceptibility to affective illness is inherited. Studies of monozygotic twins have shown that when one twin manifests the illness it is seen frequently in the co-twin. The risk for dizygotic co-twins is less but is still high; and first degree relatives have a higher incidence of the illness than would be expected by chance.

Second, there are two types of illness: bipolar, in which there are attacks of both mania and depression; and unipolar, where the illnesses are depressive only. Bipolar and unipolar families 'breed true' — that is, if illness is observed in another member of a patient's family, it tends to be of the type seen in the index individual. The characteristics of the two illnesses are different as regards age of onset, time between first and second attacks, ratio of incidence in males and females and the proportion of susceptible individuals manifesting the disorders. Thus, the illnesses are inherited in one of two forms and this fact in itself is sufficient to show that affective illnesses are biochemical in origin.

Third, instead of occurring randomly in response to traumatic life events, etc., attacks tend to follow a statistical pattern. This would not be so if they were determined predominantly by psychological factors.

Lastly, episodes of illness can be provoked, can be treated and can be prevented by physical agents. Provoking agents are reserpine, virus infections and certain hormonal disturbances. The treatments of choice are electroconvulsive therapy or antidepressant drugs: lithium salts and tricyclic antidepressants prevent recurrences of episodes.

On the basis of these data, the evidence for affective disorders being of biochemical origin is difficult to fault. Of course, this is not to say that psychologically stressful events may not play some part in precipitating or prolonging an affective illness, only that the prime mover giving rise to susceptibility to these illnesses has a biochemical basis.

The Amine Hypothesis

First attempts to understand affective disorders arose from the observations that reserpine provoked depressive illness in a proportion of presumably susceptible individuals (Lemieux, Davignon and Genest, 1958), and that reserpine depleted the brain of noradrenaline (NA) (Holzbauer and Vogt, 1956) and of 5-hydroxytryptamine (5HT) and dopamine. These two observations were put together and a causative relationship between them was assumed (Pare and Sandler, 1956). The next development from this was the catecholamine hypothesis, in which depression and mania were attributed to hypo- and hyper-function of noradrenergic pathways respectively (Kety, 1962; Schildkraut, 1965), whilst others thought of depressive illness in terms of 5HT dysfunction (Coppen, Shaw and Farrell, 1963). Both types of hypothesis postulated reduction in the amounts of NA or 5HT available in the brain and decreased activity in these pathways in depression.

These hypotheses gained greater acceptability when it was found that these amines were neurotransmitters in the regions of the brain thought to be implicated in affective illness. In the years which followed the main aminergic pathways in the brain were plotted with great skill, and this, in turn, has been succeeded by a most creditable expansion in physiological and pharmacological information on these systems. Our initial knowledge, however, derived from the first generation of antidepressant drugs, had a common factor, that these drugs shared the property of increasing the amounts either of noradrenaline alone, or of noradrenaline and 5HT, or of these two amines and dopamine, at central synapses. This seemed to support the amine hypothesis and was used by proponents of both catecholamine and indolamine versions in support of their views.

Several problems arose which made it hard to accept the initial amine hypotheses as they stood. One of these was that tricyclic antidepressants altered the behaviour of aminergic neurones by preventing reuptake of the amines across the presynaptic membrane, thus enhancing concentrations of these compounds in the synaptic cleft. This process was rapid in animals (Schildkraut, Dodge and Logue, 1969) and presumably took place with similar alacrity in humans, as evidenced by the early appearance of central side effects such as drowsiness. Despite this, recovery from depression, though progressive, was very slow, and full recovery usually took 4—8 weeks to be completed. There was a discrepancy, therefore, between the timing of the pharmacological events presumed to be therapeutic and the antidepressant response.

Further doubts about the initial amine hypotheses came when it was shown that short term depletion of NA or of 5HT did not replicate the depressive syndrome, and attempts to treat depression by potentiating catecholamine function with l-dopa, l-dopa plus a peripheral decarboxylase inhibitor, or cocaine have all failed (Mendels and Frazer, 1974; Mendels *et al*, 9175; Goodwin *et al*, 1972). The therapeutic effect of the precursor of 5HT, the amino acid tryptophan, by itself, was the subject of controversy, but it did not seem to have an early beneficial response in most severely depressed individuals (Mendels *et al*, 1975).

It was tempting to look elsewhere than at aminergic pathways for the causes of affective illness, and this may yet prove necessary. On the other hand, circumstantial evidence associating these disorders with the aminergic pathways was attracting alternative versions of the amine hypothesis. Evidence from a number of sources had assisted in this, amongst which were the following:

a) As mentioned above, there was a difference in time relationships between the rate at which tricyclic antidepressants altered aminergic systems and the speed of their therapeutic effects.

b) Many of the tricyclic antidepressants can be classified into tertiary or secondary amine compounds. The former acted mostly on 5HT neurones (Carlsson *et al*, 1969), the latter largely on NA pathways (Carlsson *et al*, 1969). Since the tertiary amine tricyclics were, in part, metabolised to secondary amine tricyclics, giving the tertiary amine compound should have resulted in increased availability of both 5HT and NA at their respective synapses. Thus the secondary amine tricyclics (and also maprotiline) (Delini-Stula, 1972) acted on noradrenergic synapses and at the same time were effective anti-depressants. It has been assumed that these two phenomena were related (i.e. their antidepressant action was due to their ability to enhance noradrenergic activity). This may be true, but it has yet to be proved.

c) In animals, increasing serotoninergic function diminished food intake and sexual activity (Gessa and Tagliamonte, 1974; Blundell, 1977). In man, loss of appetite and diminished sex drive were prominent symptoms of depression, so unless man is very different from several mammalian species, a decrease in 5HT function in depression was most unlikely.

d) Imipramine was a tertiary amine tricyclic antidepressant which therefore would increase the availability of both 5HT and NA. Patients treated with imipramine continued to get better when the synthesis of noradrenaline was prevented, but they deteriorated when the production of 5HT was blocked (Shopsin et al, 1975). This investigation suggested that the 5HT pathway was involved in the recovery process and that its integrity was essential to the establishment of normal mood.

e) Monoamineoxidase inhibitors (MAOI) increased the amounts of 5HT, noradrenaline and dopamine in their respective neurones. If, however, the 5HT neurones were 'favoured' by giving l-tryptophan, the rate of recovery was increased (Coppen, Shaw and Farrell, 1963). This probably means that 'overflow conditions' at 5HT synapses (i.e. providing an excess of this transmitter amine) potentiated the return to normal mood.

f) Two drugs, pizotifen and mianserin, have 5HT receptor-blocking activity (Speight and Avery, 1972; Kopera, 1975). Of these two drugs, mianserin was an established antidepressant (Kopera, 1975). Good rates of recovery from depression were seen during treatment with pizotifen and maprotiline (Shaw et al, 1977) and depressive symptoms have been observed on withdrawal of pizotifen (Hughes and Foster, 1971).

g) The dose response curves seen with two secondary amine tricyclics, nortriptyline and protriptyline, were in the form of an inverted 'U' (Asberg et al, 1971; Kragh-Sorensen et al, 1973; Kragh-Sorensen et al, 1976; Whyte et al, 1976). In other words, low and high plasma levels gave inferior results and good responses were only obtained with intermediate plasma concentrations. This was in contrast with two tertiary amine tricyclic antidepressants, in which the therapeutic response was proportional to plasma level, and there was little indication of an upper limit to the useful concentration of the drugs in plasma (Braithwaite *et al*, 1973; Gram *et al*, 1976; Galssman *et al*, 1977). One of the possible explanations for these findings was that both types of tricyclic drug increased the amounts of amines at their respective synapses, but that secondary amine tricyclics were therapeutically effective in doses which enhanced noradrenergic activity. In contrast, the tertiary amine tricyclic compounds, in this hypothesis, were increasing the amounts of 5HT at already 'overactive' serotoninergic synapses, and either inducing tachyphylaxis or, perhaps more likely, allowing 5HT supersensitivity to subside.

Thus, the most economical explanation for these observations was to think of depression in terms of serotoninergic overactivity. In this view, tricyclic and other antidepressant drugs would be producing different effects on noradrenergic and serotoninergic neurones, and the immediate causal abnormality in unipolar illness would be in the 5HT pathways. Of the points used in this argument, d) probably will not be replicated for ethical reasons, f) may need to be reconsidered when we know more about central 5HT receptors and g) probably needs more extensive studies before the current findings are regarded as fully established.

A complete assessment of the hypothesis, therefore, must await advances in several areas of investigation, but the hypothesis is in accord with the data available at the time of writing.

References

Asberg, M., Cronholm, B., Sjoquist, F. and Tuck, D. (1971). *Brit. Med. J.* **3**, 331.

Blundell, J. (1977). Communication to joint meeting of British Association for Psychopharmacology and Society for Studying Obesity, on 1st April, 1977.

Braithwaite, R.A., Goulding, R. Theano, G., Bailey,J . and Coppen, A. (1973). *Lancet* i, 556.

Carlsson, A., Corrodi, H., Fuxe, K. and Hokfelt, T. (1969). *Europ. J. Pharmacol.* **5**, 357.

Carlsson, A., Corrodi, H., Fuxe, K. and Hokfelt, T. (1969). *Europ. J. Pharmacol.* **6**, 367.

Coppen, A., Shaw, D.M. and Farrell, J.P. (1963). *Lancet* i, 79.

Delini-Stula, A. (1972). *In* "Depressive Illness" (ed. Kielholz, P.), p. 113. Hans Huber Publishers, Bern.

Gessa, G.L. and Tagliamonte, A. (1974). *Adv. Biochem. Psychopharmac.* **11**, 217.

Glassman, A.H., Perel, J.M., Shostak, M., Kantor, S.J. and Fleiss, J.L. (1977). *Arch. Gen. Psychiat.* **34**, 197.

Goodwin, F.K., Post, R.M. and Kotin, J. (1972). *Psychopharmacologia* **26**, Suppl. 82.

Gram, L.F., Reisby, N., Ibsen, I., Nagy, A., Dencker, S.J., Beck, P., Petersen, G.O. and Christiansen, J. (1976). *Clin. Pharmacol. Therapeut.* **19**, 318.

Holzbauer, M. and Vogt, M. (1956). *J. Neurochem.* **1**, 8.

Hughes, R.C. and Foster, J.B. (1971). *Curr. Therap. Res.* **13**, 63.

Kety, S.S. (1962). *In* "Ultrastructure and metabolism of the nervous system" (eds. Korey, S.R., Pope, A. and Rodins, A.), p. 311, Vol. 40. *Res. Publ. Ass. Neuro. Ment. Dis.*, Williams and Wilkins Co., Baltimore.

Kopera, H. (1975). *In* "Depressive Illness and Experiences with a new anti-depressant drug GB94" (ed. Vossenaar, T.), p. 26. Excerpta Medica, Amsterdam.

Kragh-Sorenson, P., Asberg, M. and Hansen, C.E. (1973). *Lancet* **i**, 113.

Kragh-Sorensen, P., Hansen, C.E., Baastrup, P.C. and Hvidberg, E.F. (1976). *Pharmakopsych.* **9**, 27.

Lemieux, G., Davignon, A. and Genest, J. (1958). *Amer. J. Psychiat.* **115**, 459.

Mendels, J. and Frazer, A. (1974). *Arch. Gen. Psychiat.* **30**, 447.

Mendels, J., Stinnett, J.L., Burns, D. and Frazer, A. (1975). *Arch. Gen. Psychiat.* **32**, 22.

Pare, C.M.B. and Sandler, M. (1956). *J. Neurol. Neurosurg. Psychiat.* **22**, 247.

Schildkraut, J.J. (1965). *Amer. J. Psychiat.* **122**, 509.

Schildkraut, J.J., Dodge, G.A. and Logue, M.A. (1969). *J. Psychiat. Rev.* **7**, 29.

Shaw, D.M., Riley, G., Michalakeas, A.C., Tidmarsh, S.F., Blazek, R. and Johnson, A.L. (1977). *Lancet* **i**, 1259.

Shopsin, B., Gershon, S., Goldstein, M., Friedman, E. and Wilk, S. (1975). *Psychopharmac. Commun.* **1**, 239.

Speight, T.M. and Avery, G.S. (1972). *Drugs* **3**, 159.

Whyte, S.F., MacDonald, A.J., Naylor, G.J. and Moody, J.P. (1976). *Brit. J. Psychiat.* **128**, 384.

DOPAMINERGIC PROCESSES IN SCHIZOPHRENIA AND THE MECHANISM OF THE ANTIPSYCHOTIC EFFECT

T.J. CROW and EVE C. JOHNSTONE

*Division of Psychiatry, Clinical Research Centre,
Northwick Park Hospital, Harrow, Middlesex, U.K.*

Introduction

The similarity of amphetamine psychosis and acute paranoid schizo-phrenia (Connell, 1958), together with observations on the mechanism of action of the amphetamines (Randrup and Munkvad, 1967) and investigations of the mode of action of antipsychotic drugs, have led to interest in the possible role of dopamine in schizophrenia. These two strands of thought come together in experiments (Randrup and Munkvad, 1965) demonstrating selective reversal of the behavioural effects of amphetamine by neuroleptic drugs. From such observations developed the 'dopamine hypothesis' of schizophrenia — that some or all of the symptoms may result from excessive activity of dopaminergic mechanisms, and that the adverse effects of this activity are reduced by dopamine receptor blockade induced by neuroleptic drugs.

Amphetamine Psychosis

In his classic monograph Connell (1958) drew attention to the fact that many of the most characteristic symptoms of acute paranoid schizophrenia are seen in patients who have received excessive doses of amphetamines, but who do not otherwise appear to suffer from schizophrenia. Ellinwood (1967) confirmed and extended these observations on amphetamine addicts.

The fact that psychotic changes can be observed in most, if not all, experimental subjects if sufficiently large doses are administered (Griffith, Cavanaugh, Held and Oates, 1972) makes it unlikely that this is an idiosyncratic reaction which occurs only in subjects pre-disposed to schizophrenia, and the speed of onset is sometimes such as to rule out the possibility that the changes are secondary to sleep deprivation.

In animal experiments many amphetamine-induced behavioural changes have been shown to be dependent upon dopamine release. In

rats stereotyped sniffing, licking and gnawing behaviours are abolished
by drugs which deplete both dopamine and noradrenaline but not by
those which deplete noradrenaline alone (Randrup and Munkvad,
1966), and this syndrome is also abolished by destruction of the
major ascending dopaminergic projections by localised intracerebral
injections of 6-OH-dopamine (Creese and Iversen, 1975). Such observations suggest the possibility that the amphetamine psychosis is
also dependent upon increased dopamine release. CSF sampling
techniques have provided some evidence (Angrist, Sathananthan, Wilk
and Gershon, 1974) that psychotogenic doses do increase dopamine

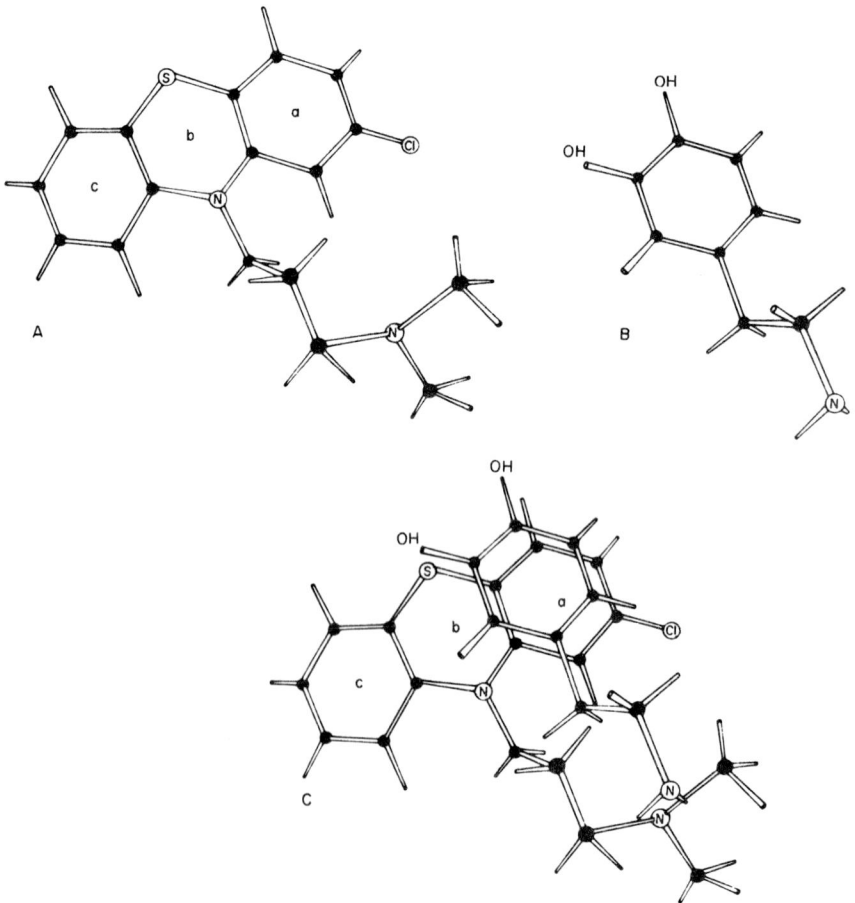

Fig. 1 The three-dimensional structures of chlorpromazine (A), dopamine (B) and their possible inter-relation (according to Horn and Snyder, 1971).

turnover, and some at least of the psychological changes induced by the amphetamines can be reversed by drugs such as α-methyl-p-tyrosine, pimozide (Gunne, Anggard and Jonsson, 1972) and halo-peridol (Angrist, Lee and Gershon, 1974) which interfere with dopaminergic transmission.

The symptoms of acute schizophrenia are exacerbated by methylphenidate and amphetamine (Janowsky, El Yousef, Davis and Sekerke, 1973) but chronic schizophrenic patients appear to be relatively insensitive to the amphetamines (Kornetsky, 1976). Thus it seems possible that in the chronic state, or perhaps in some subgroup of illnesses with a deteriorating course, the state of dopaminergic mechanisms may be quite different from that seen in acute psychoses.

Neuroleptics and Dopamine Receptor Blockade

Neuroleptic drugs increase dopamine turnover (Carlsson and Lindqvist, 1963; O'Keefe, Sharman and Vogt, 1970) and this action has been interpreted as a feedback response of the presynaptic neurone to

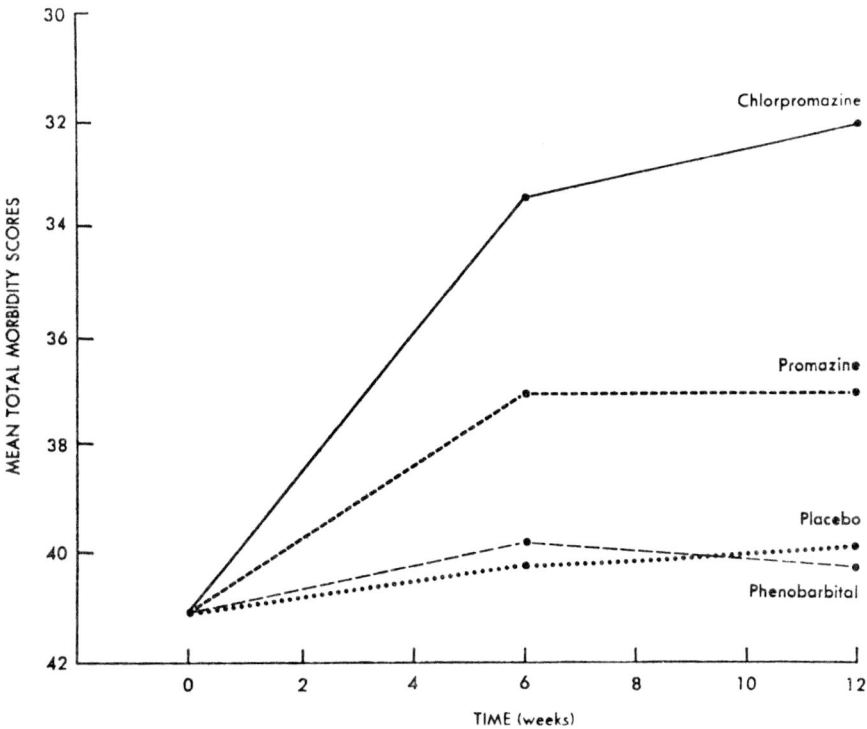

Fig. 2 The effects of chlorpromazine, promazine and phenobarbital in acute schizophrenia (reproduced from Klein and Davis, 1969). Data taken from VA studies I and II (39, 40).

blockade of the postsynaptic receptor. Stereochemical considerations (Horn and Snyder, 1971) suggest that the conformation of chlorpromazine may be complementary to that of dopamine (Fig. 1) but that this relationship depends upon the structural rigidity imparted to the dimethyl-amino-propyl side chain by the presence of the chlorine substituent. Thus promazine, which lacks the chlorine atom but otherwise resembles chlorpromazine, would be expected to be a less active dopamine receptor antagonist. In clinical practice promazine is less effective in ameliorating the symptoms of schizophrenia than chlorpromazine, although more effective than phenobarbital (Fig. 2).

The development of in vitro assays of the dopamine receptor has facilitated investigations of the structural requirements for dopamine receptor blockade. For example, a dopamine-sensitive adenylate cyclase has been described in corpus striatum. In a series of compounds of the phenothiazine and thiaxanthene groups there is a strong positive relationship between dopamine blocking activity tested in this system and antipyschotic potency (Miller, Horn and Iversen, 1974). Haloperidol-binding has also been suggested as an index of the dopamine receptor and there is a high correlation between antipsychotic activity and inhibition of haloperidol-binding (Creese, Burt and Snyder, 1976; Seeman, Lee, Chau-Wong and Wong, 1976).

An Experimental Test of the Dopamine Blockade Hypothesis

A direct clinical test of the dopamine blockade hypothesis was made possible by the discovery (Miller, Horn and Iversen, 1974) that the stereoisomers of certain thiaxanthene compounds differ widely in their ability to block the dopamine receptor. In particular the α- (or cis-) isomer of the widely used neuroleptic flupenthixol (Fig. 3) is more than a thousand times more potent than the β- (or trans-) isomer.

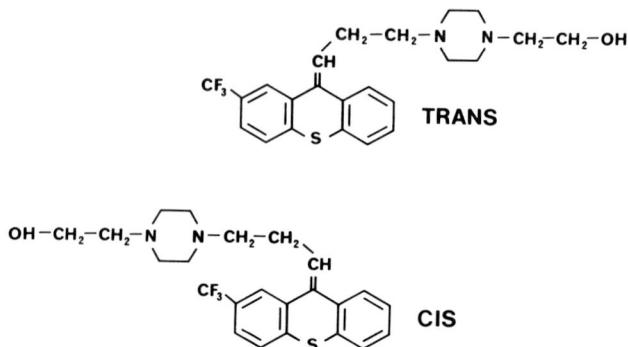

Fig. 3   The α- (cis-) and β- (trans-) isomers of flupenthixol.

The dopamine blockade hypothesis clearly predicts that only the α-isomer should be active clinically. If the β-isomer were equally active this would be a decisive refutation. If the β-isomer were shown to have some antipsychotic activity, even though less than α-flupenthixol, this might be interpreted as evidence that actions other than dopamine receptor blockade are also relevant to the therapeutic effect.

In 45 patients with acute schizophrenic illnesses (as defined by Present State Examination criteria) of recent onset, α-flupenthixol was found to be significantly more effective than the β-isomer, which itself was marginally but not significantly less effective than placebo (Fig. 4). The findings are consistent with the hypothesis that dopamine

Flupenthixol trial

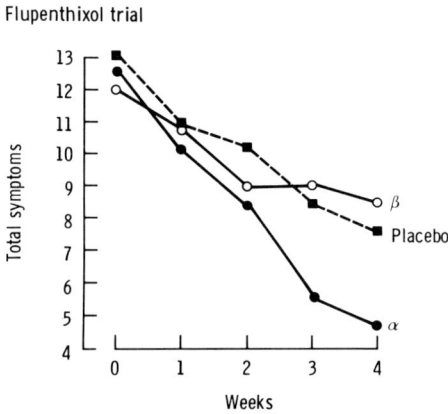

Fig. 4 Clinical ratings (total symptoms) in patients on α- and β-flupenthixol and placebo over four weeks of treatment.

receptor blockade is the sole component of the antipsychotic effect.

However, other possible mechanisms can only be ruled out if it is known that the two isomers do not differ with respect to these actions. Such knowledge is available for certain actions (Table I), but it is apparent that the α-isomer is probably considerably more potent than the β-isomer as a serotonin antagonist. However, within a series of neuroleptics interaction with the serotonin receptor is poorly correlated with therapeutic activity (Bennett and Snyder, 1975). It may be concluded that the dopamine blockade hypothesis has survived a stringent test, and that on the basis of present knowledge dopamine receptor antagonism is the only predictor of antipsychotic potency.

Analysis of individual symptom ratings (Fig. 5) revealed that the relative superiority of α-flupenthixol over the β-isomer and placebo was seen almost exclusively on what may be referred to as the

TABLE I

Relative Potencies of the Two Isomers of Flupenthixol

$\alpha = \beta$ $\alpha \cong \beta$	$\alpha \gg \beta$

noradrenaline receptor:
   inhibition of noradrenaline-
   sensitive adenylate cyclase
 	
 cholinergic receptor:
   inhibition of QNB-binding and
   ACh antagonism
 	
 opiate receptor:
   inhibition of naloxone and
   dihydromorphine-binding
 	
 inhibition of dopamine uptake
 	
 inhibition of GABA uptake
	dopamine receptor:
	   inhibition of dopamine or
	   haloperidol binding, inhibition
	   of dopamine-sensitive adenylate
	   cyclase
	
	 serotonin receptor:
	   inhibition of 5-HT and LSD binding
	   5-HT antagonism

For data and references see: Miller, Horn and Iversen, 1974; Moller-Nielsen *et al*, 1973; Horn and Phillipson, 1975; Enna *et al*, 1976.

positive symptoms (delusions, hallucinations, thought disorder and incongruity of affect) and was much less apparent in the non-psychotic (depression, anxiety and retardation) and negative (flattening of affect, mutism) features of the disease. Thus if dopamine receptor blockade is relevant to the antipsychotic effect it appears to affect positive and not negative schizophrenic symptoms.

Subdivision of the patient sample by different diagnostic criteria

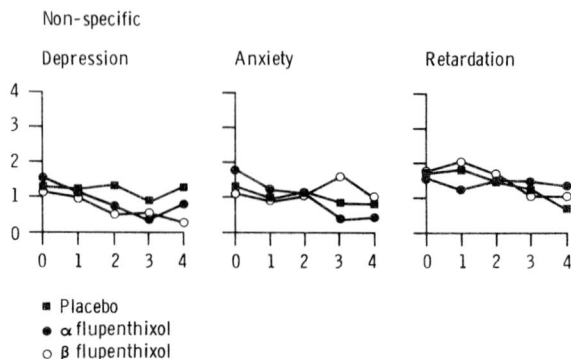

■ Placebo
● α flupenthixol
○ β flupenthixol

Fig. 5a. Effects of α- (cis-) and β- (trans-) flupenthixol and placebo on the non-specific symptoms. Abscissa — weeks of treatment.

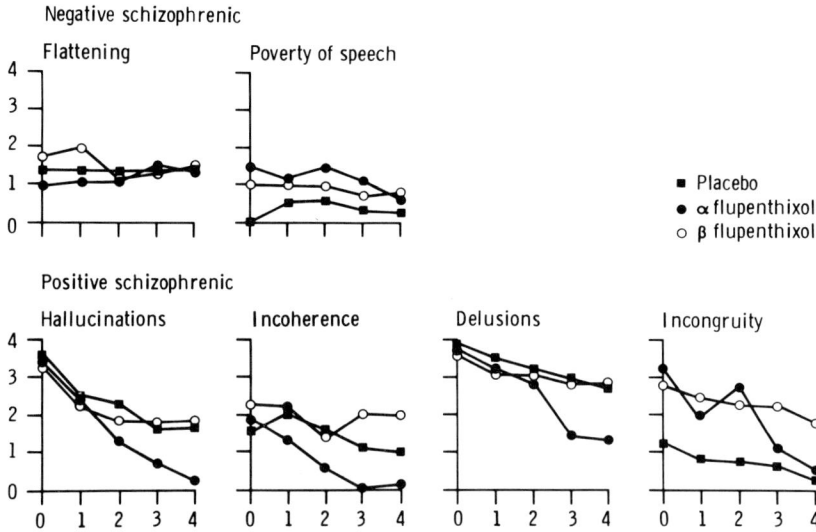

Fig. 5b. Effects of α- (cis-) and β- (trans-) flupenthixol and placebo on the negative and positive schizophrenic symptoms. Abscissa — weeks of treatment.

provided some information on the type of schizophrenic illness most susceptible to the drug effect. The total sample of 45 patients was subdivided first by the Feighner criteria (which stipulate an element of progressive deterioration) and secondly by the presence or absence of affective features (i.e. schizoaffective vs. non-schizoaffective). In each case calculation of the improvement in patients on α-flupenthixol minus the mean improvement on the β-isomer or placebo gave an estimate of the magnitude of the effect which could be attributed to drugs. The results (Fig. 6) show that the antipsychotic effect is greater in Feighner positive and non-schizoaffective patients than in the remaining patients, i.e. that the effect is greatest in the most typically schizophrenic illnesses.

The Time Course of the Antipsychotic Effect

The tubero-infundibular dopamine system in the median eminence of the hypothalamus inhibits prolactin release. The increase in prolactin secretion which occurs in patients on neuroleptic drugs can therefore be used as an independent index of dopaminergic blockade. In the above trial prolactin was found to increase in patients on α- but not β-flupenthixol (Cotes, Crow and Johnstone, 1977). By subtracting the weekly ratings of patients on the α-isomer from the mean ratings of those on the β-isomer and placebo it was possible to compare the

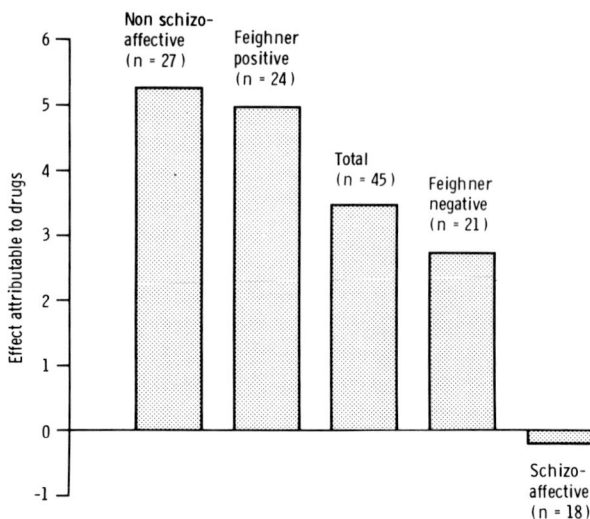

Fig. 6 Improvement attributable to drugs (calculated as change in total ratings on α-flupenthixol minus mean change on β-flupenthixol and placebo) in patients with and without affective symptoms (schizoaffective and non-schizoaffective patients respecively) and in those with and without evidence of deterioration. (Feighner positive and negative). The drug effect is greater in non-schizoaffective and Feighner positive patients than in the remaining group in each case.

time course of dopamine receptor blockade, as indicated by prolactin release, with that of the therapeutic effects of drug administration (Fig. 7).

There is substantial discrepancy. Whereas the increase in prolactin secretion is well established at the end of the first week, and may well be present long before this, the therapeutic effect is greatest in the third week. Thus there is a temporal dissociation between dopamine receptor blockade and the antipsychotic effect. It is as if dopamine receptor blockade is a necessary condition for some other, and longer term, change to take place which is itself reflected in the clinical state.

The Site of the Antipsychotic Action

A stumbling block to the view that dopaminergic blockade is the mechanism of the antipsychotic effect has been the fact that extra-pyramidal effects are not closely related to antischizophrenic action (NIMH Psychopharmacology Service Center, 1964). For example, thioridazine is approximately equipotent to chlorpromazine in acute schizophrenia but induces fewer parkinsonian effects. Miller and Hiley (1974) demonstrated that antimuscarinic potency may be

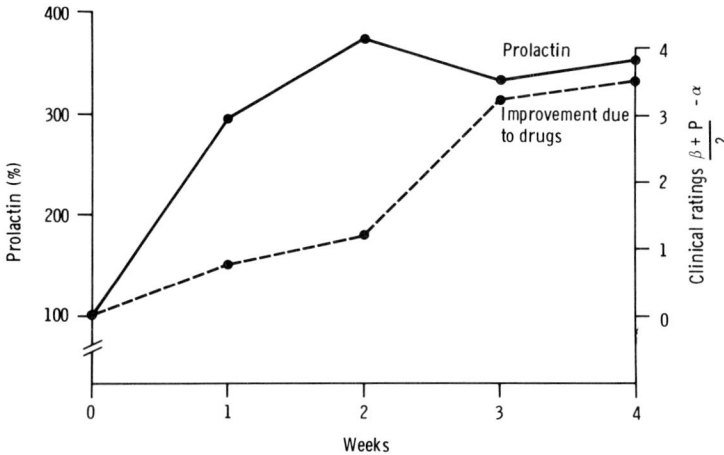

Fig. 7 Time course of the rise in prolactin compared with the time course of the therapeutic effect of drug treatment.

relevant to extrapyramidal actions, drugs such as thioridazine having a high affinity for the cholinergic receptor and perhaps for this reason possessing 'in-built' anti-parkinsonian activity. However, the antipsychotic actions appear not to be modified by muscarinic antagonist activity, and therefore it must follow that antipsychotic and extrapyramidal effects take place at different sites within the C.N.S. Anden (1972) suggested that the antipsychotic effect might be due to dopamine receptor blockade occurring at a site within the mesolimbic system, e.g. the nucleus accumbens, while the extra-pyramidal effects occurred within the corpus striatum.

A recent comparison (Crow, Deakin and Longden, 1977) of equivalent therapeutic doses of the three drugs thioridazine, chlorpromazine and fluphenazine established that the effects on dopamine turnover in the striatum, assessed by homovanillic acid (HVA) accumulation, were in the rank order of the drugs' ability to induce extrapyramidal effects, while in the nucleus accumbens the effects were closely related to antipsychotic potency (Fig. 8).

Thus while the antipsychotic effect cannot occur by dopaminergic blockade within the corpus striatum a site of action within the nucleus accumbens is entirely possible. The possibility of an action at sites elsewhere within the mesolimbic system (e.g. at dopaminer-gically innervated sites in the frontal cortex) remains to be assessed.

Are Dopamine Neurones Overactive in Schizophrenia?

There are now a number of observations which suggest either that

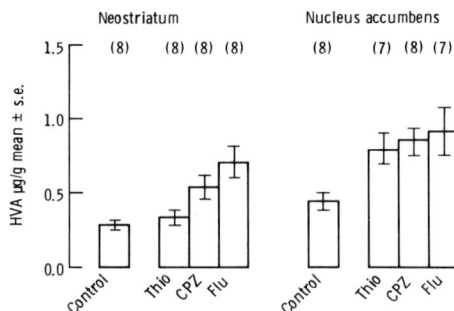

Fig. 8   Effects of thioridazine (Thio), chlorpromazine (CPZ) and fluphenazine (Flu) on homovanillic acid (HVA) accumulation in rat neostriatum and nucleus accumbens. The dose ratios (1:1:0.05) were selected to correspond to therapeutically equivalent daily dosages (National Institute of Mental Health −PSC, 1964). Figures in parentheses represent numbers of animals studied.

dopamine neurones are not overactive in schizophrenia, or at least that this cannot be the primary disturbance:

a) Studies of HVA levels in CSF using the probenecid technique (Bowers, 1974; Post, Fink, Carpenter and Goodwin, 1975) have failed to reveal evidence of increased dopamine turnover, although such an increase has been demonstrated in this way in amphetamine psychosis (Angrist, Sathananthan, Wilk and Gershon, 1974). In schizophrenia there is even evidence that with increasing severity there is a decrease in turnover (Bowers, 1974; Post *et al*, 1975).

b) Prolactin secretion is not decreased in either acute (Meltzer, Sachar and Frantz, 1974) or chronic (Johnstone, Crow and Mashiter, 1977) unmedicated patients. Thus there is no evidence of increased dopamine release from the tubero-infundibular system.

c) Schizophrenia-like illnesses can occur in patients with long-standing Parkinson's disease (Crow, Johnstone and McClelland, 1976) and the symptoms of either illness do not appear to be substantially modified by the presence of the other. Since dopamine is as depleted in mesolimbic as in striatal regions in Parkinson's disease (Farley, Price and Hornykiewicz, 1976) one may conclude that increased dopamine release is not necessary for schizophrenic symptoms to occur.

d) In postmortem specimens HVA concentrations are not significantly different in the brains of controls and patients who have suffered from schizophrenia, either in the striatum or nucleus accumbens (Crow, Johnstone, Longden and Owen, in press).

For these reasons it seems unlikely that a simple increase in dopamine release is the pathogenic mechanism in the majority of schizophrenic illnesses.

Conclusions

The dopamine blockade hypothesis of the antipsychotic effect can explain the efficacy of a wide range of neuroleptic drugs and has survived a stringent clinical test. The antipsychotic effect occurs in the most typically schizophrenic illnesses and upon characteristically schizoprenic symptoms, although it may not be relevant to the negative features which are so important in the development of the defect state. This finding, together with observations on the effects of amphetamine in acute and chronic schizophrenic illnesses, suggests that different mechanisms may be operative in the chronic state, and that dopamine receptor blockade may be more relevant to the acute than the chronic condition. In acute schizophrenia the time course of the effect is such that it seems likely that the therapeutic effects are an indirect and delayed consequence of dopamine receptor blockade. The evidence is consistent with the hypothesis that the antipsychotic effect is secondary to dopamine receptor blockade occurring within the nucleus accumbens.

Three independent types of observation have cast doubt on the

Hypothesis I - decreased activity of a system
            antagonistic to the mesolimbic
            dopamine system

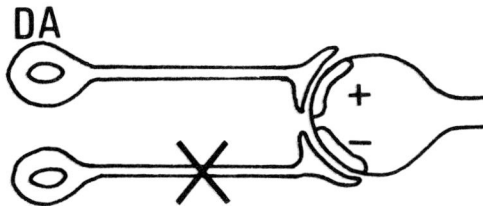

Hypothesis II - supersensitivity of mesolimbic
            dopamine receptors

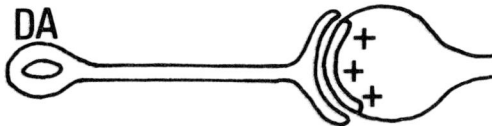

Fig. 9 Two alternatives to the dopamine neurone overactivity hypothesis of schizophrenia.

concept of dopamine neurone overactivity as the primary defect in schizophrenia. Two alternative hypotheses (Fig. 9) are that there may be supersensitivity of the dopamine receptor, or some neural element distal to the receptor, or that there may be a deficit in a system which acts in antagonism to the mesolimbic dopamine system. In either case a reduction in the effectiveness of dopaminergic transmission might return the total system toward normality.

References

Anden, N-E. (1972). *In* "Mechanism of release of biogenic amines" (eds. von Euler, U.S., Rosell, S. and Uvnas, B.), 357. Pergamon, Oxford.

Angrist, B., Lee, H.K. and Gershon, S. (1974). *Am. J. Psychiat.* **131**, 817–819.

Angrist, B., Sathananthan, G., Wilk, S. and Gershon, S. (1974). *J. Psychiat. Res.* **11**, 13–23.

Bennett, J.P. and Snyder, S.H. (1975). *Brain Res.* **94**, 523–544.

Bowers, M.B. (1974). *Arch. Gen. Psychiat.* **31**, 50–54.

Carlsson, A. and Lindqvist, M. (1963). *Acta. Pharm. Toxicol.* **20**, 140–144.

Connell, P.H. (1958). Amphetamine Psychosis. Oxford University Press, London.

Cotes, P.M., Crow, T.J. and Johnstone, E.C. (1977). *Brit. J. Clin. Pharmacol.* (in press).

Creese, I., Burt, D.R. and Snyder, S.H. (1976). *Science* **192**, 481–483.

Creese, I. and Iversen, S.D. (1975). *Brain Res.* **83**, 419–436.

Crow, T.J., Deakin, J.F.W. and Longden, A. (1977). *Psychol. Med.* **7**, 213–221.

Crow, T.J., Johnstone, E.C., Longden, A. and Owen, F. (1977). *In* "Proceedings of International Society of Neurochemistry Conference in Dopamine" (in press).

Crow, T.J., Johnstone, E.C. and McClelland, H.A. (1976). *Psychol. Med.* **6**, 227–233.

Ellinwood, E.H. (1967). *J. Nerv. Ment. Dis.* **144**, 274–283.

Enna, S.J., Bennett, J.P., Burt, D.R., Creese, I. and Snyder, S.H. (1976). *Nature (Lond.)* **263**, 338–347.

Farley, I.J., Price, K.S. and Hornykiewicz, O. (1977). *In* "Non-striatal dopamine" (eds. Costa, E. and Gessa, G.L.). Raven Press, New York.

Griffith, J.D., Cavanaugh, J., Held, J. and Oates, J.A. (1972). *Arch. Gen. Psychiat.* **26**, 97–100.

Gunne, L.M., Angaard, E. and Jonsson, L.E. (1972). *Psychiat. Neurol. Neurochirurgia (Amsterdam)* **75**, 225–226.

Horn, A.S. and Phillipson, O.T. (1975). *Brit. J. Pharmacol.* **55**, 299–300P.

Horn, A.S. and Snyder, S.H. (1971). *Proc. Nat. Acad. Sci.* **68**, 2325–2328.

Janowsky, D.S., El-Yousef, M.K., Davis, J.M. and Sekerke, H.J. (1973). *Arch. Gen. Psychiat.* **28**, 185–191.

Johnstone, E.C., Crow, T.J. and Mashiter, K. (1977). *Psychol. Med.* **7**, 223–228.

Klein, D.F. and Davis, J.M. (1969). Diagnosis and Drug Treatment of Psychiatric Disorders. Williams and Wilkins, Baltimore.

Kornetsky, C. (1976). *Arch. Gen. Psychiat.* **33**, 1425–1428.

Meltzer, H.Y., Sachar, E.J. and Frantz, A.G. (1974). *Arch. Gen. Psychiat.* **31**, 564–569.

Miller, R.J. and Hiley, C.R. (1974). *Nature (Lond.)* **248**, 596–597.

Miller, R.J., Horn , A.S. and Iversen, L.L. (1974). *Molec. Pharmacol.* **10**, 759–766.

Moller-Nielsen, I., Pedersen, V., Nymark, M., Franck, K.F., Boeck, V., Fjalland, B. and Christensen, A.V. (1973). *Acta pharmacol. et toxicol* **33**, 353–362.

NIMH-PSC (1964). *Arch. Gen. Psychiat.* **10**, 246–261.

O'Keefe, R., Sharman, D.R. and Vogt, M. (1970). *Brit. J. Pharmacol.* **38**, 287–304.

Post, R.M., Fink, E., Carpenter, W.T. and Goodwin, F.K. (1975). *Arch. Gen. Psychiat.* **32**, 1013–1069.

Randrup. A. and Munkvad, I. (1965). *Psychopharmacologia* **7**, 416–422.

Randrup, A. and Munkvad, I. (1966). *Nature (Lond.)* **211**, 540.

Randrup, A. and Munkvad, I. (1967). *Psychopharmacologia* **11**, 300–310.

Seeman, P., Lee, T., Chau-Wong, M. and Wong, K. (1976). *Nature (Lond.)* **261**, 717–719.

APPROACHES TO THE STUDY OF PEPTIDES ENDOGENOUS TO THE BRAIN

CHRISTOPHER R. SNELL

Laboratory of Peptide Chemistry, National Institute for Medical Research, Mill Hill, London, U.K.

Introduction

Initial impetus to the study of peptides in brain came in the sixties, with the discovery in the hypothalamus of peptides secreted into the hypothalamo-hypophyseal portal system to control the release of pituitary hormones into the general circulation. However, in the last five years attention has increasingly focussed on the effects of peptides within the central nervous system acting, not as neurohormones, but as neurotransmitters or neuromodulators from peptidergic neurons (Zetler, 1976; Martin *et al*, 1975) involved in the maintenance of mood and behaviour. The consequent academic and clinical implications have made the isolation and elucidation of the physiological role of CNS peptides of prime importance.

Table I lists the peptides so far identified in the CNS; however, of these only a few have been subject to any rigorous examination. Comprehensive evaluation of a potential neuroactive peptide involves its unambiguous *identification* and *localisation* in neural tissue, and examination of its physiological, pharmacological and biochemical *effects* and its *degradation*. Figure 1 shows diagrammatically the

TABLE I

Peptides Identified in the CNS

Substance P	Vasopressin
Thyroliberin	Oxytocin
Luliberin	Neurotensin
Somatostatin	VIP*
Enkephalin, α and γ endorphin	CCK*
C-fragment (β-endorphin)	Gastrin*
	Sleep peptide

*Identified only by immunohistofluorescence.

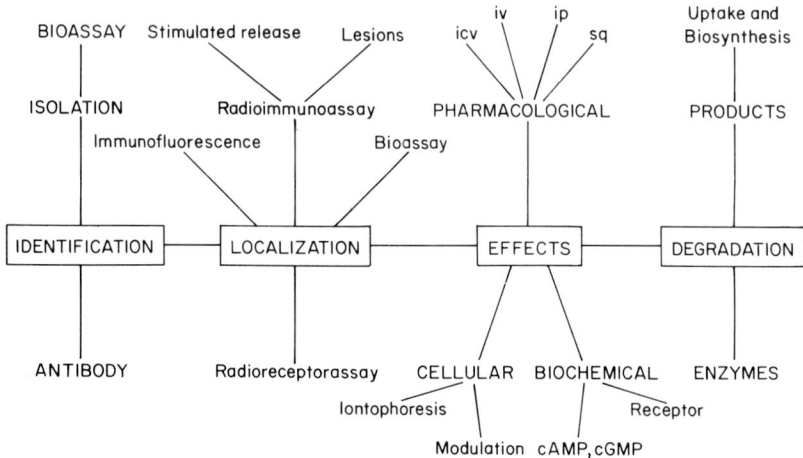

Fig. 1   Approaches to the study of neuroactive peptides.

approaches available to study these four fundamental aspects.
The identification of a peptide ideally should come from chemical
characterisation, after isolation with the aid of a suitable bioassay.
Immunological methods have also been used to deduce the presence
in brain of peptide hormones already isolated from other tissue, but
such identification can only be regarded as tentative since immunoreactivity is not sufficient proof of chemical identity. As the levels of
peptide in the brain are low, rapid evaluation of the pharmacology
and regional distribution can only follow if the peptide can be
synthesised. In the absence of a radioimmunoassay the regional
distribution can be studied using a bioassay or, if available, a radioreceptor assay. If an antibody can be developed, radioimmunoassay
can be complemented by immunohistofluorescence to identify cell
bodies and nerve terminals of the relevant peptidergic neurons.

The central effects of the peptide can be studied a) *pharmacologically* after injection, principally into the brain ventricles, but
also at peripheral sites; b) *cellularly*, both electrophysiologically using
microiontophoretic techniques, and by study of the modulatory
effects of the peptide on other neurotransmitter levels and turnover;
c) *biochemically* by examination of the interaction of the peptide
with its physiological receptor and the effect of such interactions on
intracellular cGMP and cAMP levels.

Finally, the degradation of the peptide can be studied by

identification of any intermediates formed during inactivation and by study of the enzymes responsible for such processes. A knowledge of the degradative pathway of the hormone may allow analogues to be designed that are more resistant to *in vivo* degradation; such analogues might be useful pharmacologically or even clinically.

The application of these approaches to the study of endogenous neuroactive peptides will now be described in relation specifically to the central roles of Substance P, the C-Fragment of lipotropin and the hypothalamic hypophysiotropic hormones.

Substance P

Substance P was first detected by von Euler and Gaddum (1931) in extracts of brain and intestinal muscle as a substance that stimulated smooth muscle contraction and produced peripheral vasodilation. Subsequently the sialogogic property of the peptide was used to monitor its isolation from hypothalamic tissue (Leeman and Hammerschlag, 1967; Lembeck and Starke, 1968; Chang and Leeman, 1970) and the peptide was found to have the following sequence (Chang *et al*, 1971):

Arg.Pro.Lys.Pro.Gln.Gln.Phe.Phe.Gly.Leu.Met.NH$_2$.

The development of a specific antibody (Powell *et al*, 1973) for this peptide has allowed extensive study of the regional localisation and identification of Substance P pathways in the CNS. Particularly high concentrations of Substance P were found (Brownstein *et al*, 1976; Kanazawa and Jessell, 1976) in the substantia nigra (1900 pmoles/g tissue), the interpeduncular nucleus (670 pmoles/g) and the dorsal horn of the spinal cord (1500 pmoles/g). Immunohistofluorescence methods have been used to localise Substance P immunoreactivity in the CNS (Nilson *et al*, 1974), in neuronal perikarya in dorsal root ganglia and in neuronal processes within the dorsal horn of the spinal cord (Hokfelt *et al*, 1975a, 1976). More recent ultrastructural studies have shown the spinal cord immunoreactivity to be present within vesicles at axonal termini (Pickel *et al*, 1977; Palay and Palay, 1977). Otsuka *et al* (1975) have found Substance P a powerful excitant of motor neurons in the spinal cord, with a potency some 200-fold greater than glutamic acid. In addition, levels of Substance P present in the dorsal horn decrease dramatically on ligation of the dorsal roots (Hokfelt *et al*, 1975). The excitatory effects are observed even after synaptic transmission is blocked by reduced calcium concentration in the medium (Konishi and Otsuka, 1974) or by added tetrodotoxin (Nicoll, 1976). These observations provide strong evidence that Substance P acts as a transmitter of primary afferent neurons as first suggested by Lembeck (1953). Recent iontophoretic work by Henry (1976) has shown Substance P to cause excitation only of those dorsal horn neurons that are excited by noxious radiant heat applied to the skin, suggesting that the physiological excitatory action of Substance P in the spinal cord is associated specifically with nociception.

The radioimmunoassay approach has also been used, in combination with specific brain lesions, to localise the different pathways responsible for the high levels of Substance P in the interpeduncular nucleus and the substantia nigra. The principal afferents to the interpeduncular nucleus arise from the habenular nucleus and appear to be cholinergic, but lesions of the medial habenula do lead to a decrease in levels of interpeduncular Substance P (Mroz *et al*, 1976; Emson *et al*, 1977); however, the levels do not fall below 40% of the unlesioned controls, suggesting that the habenula is not the sole source of interpeduncular Substance P. Similar studies of the levels of Substance P in the substantia nigra after specific brain lesions have shown these nerve terminals to originate from cell bodies located in the globus pallidus and corpus striatum (Emson *et al*, 1977; Mroz *et al*, 1977). The striato-nigral tract is paralleled by descending GABAnergic fibres and also by ascending dopaminergic fibres to the caudate. Interestingly, microinjection of Substance P into the substantia nigra results in contraversive circling behaviour which is accompanied by increased dopamine turnover in the striatum (Olpe and Koella, 1977; James and Starr, 1977). It would seem that the striatal-nigral Substance P neurons might act physiologically to control striatal dopamine levels, probably under the inhibitory control of the GABAnergic fibres, as *in vitro* GABA has been shown to inhibit the stimulated release of Substance P from the substantia nigra.

Substance P release from hypothalamic synaptosomes and slices has been stimulated by high K^+ concentrations (Schenker *et al*, 1976; Iversen *et al*, 1976) and by similar stimulation of dorsal roots in isolated rat spinal cord (Otsuka and Konishi, 1976). All of this is strong evidence for a neurotransmitter function for Substance P.

The degradation and biosynthesis of Substance P have received little attention; nevertheless the physiological and pharmacological evidence makes this the most established of all the potential peptide neurotransmitters.

C-Fragment of lipotropin (residues 61−91; β-endorphin)

The initial isolation and identification of the morphinomimetic peptide methionine enkephalin (H.Tyr.Gly.Gly.Phe.Met.OH) by

Hughes *et al* (1975) has stimulated intensive research on many aspects of endogenous opiate peptides. Figure 2 shows diagrammatically the structural relationship between the opiate peptides within the 61–91 region of lipotropin.

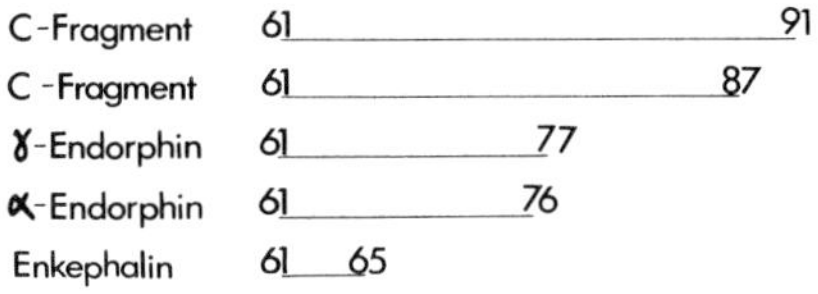

C-Fragment	61	91
C -Fragment	61	87
ɣ-Endorphin	61	77
⍺-Endorphin	61	76
Enkephalin	61	65

Fig. 2 Sequences of opiate peptides related to C-Fragment. Residue numbers refer to position in β-lipotropin sequence.

In addition, an enkephalin analogue with leucine in place of methionine has also been reported (Hughes *et al*, 1975; Simantov and Snyder, 1976). The C-Fragment of lipotropin was first isolated from pituitary (Bradbury *et al*, 1975; Li and Chung, 1976; Graf *et al*, 1976) and has since been found in brain using radioreceptor assay (Bradbury *et al*, 1976a) and radiommunoassay (Krieger *et al*, 1977; Snell and Smyth, unpublished). C'-Fragment, the first 27 residues of C-Fragment, was also isolated from pituitary tissue (Bradbury *et al*, 1975). α and γ endorphin were isolated from hypothalamo-pituitary tissue (Gullemin *et al*, 1976), and correspond to the first 16 and 17 residues of C-Fragment respectively. Methionine enkephalin is identical with the aminoterminal pentapeptide of C-Fragment.

Some pharmacological effects of C-Fragment on intraventricular injection are listed in Table II. On a molar basis the peptide is 50–100 times more potent than morphine in producing these effects, and it is clearly acting at the opiate receptor, as the actions are reversible by the opiate antagonist, naloxone. The shorter peptides, on the other hand, produce these effects transiently or not at all, even at much higher doses (Jacquet and Marks, 1976; Bloom *et al*, 1976; Buscher *et al*, 1976; Belluzzi *et al*, 1976; Feldberg and Smyth, 1977). Modified forms of the pentapeptide have successfully been developed that can produce analgesia (Pert *et al*, 1976; Coy *et al*, 1976; Bradbury *et al*, 1976; Bajusz *et al*, 1977; Roemer *et al*, 1977), catatonia (Roemer *et al*, 1977) and release of prolactin and growth hormone (Cusan *et al*, 1977). The most potent of these modified peptides (Roemer *et al*, 1977) produces analgesia on intraventricular injection at molar doses similar to that of C-Fragment.

At the biochemical level, C-Fragment inhibits prostaglandin E_1 stimulated cAMP production in rat brain homogenates (Collier and Roy, 1977) as a consequence of receptor interaction. Lampert *et al*,

TABLE II

Central Pharmacological Effects of C-Fragment on Intraventricular Injection

Effect	Dose (nmoles)	Species	Reference
analgesia and catalepsy	3	cat	Feldberg and Smyth, 1976
analgesia	0.3	rat	Graf *et al*, 1976; Loh *et al*, 1976b; Van Ree *et al* 1976; Bradbury *et al*, 1977
hyperglycaemia	20	cat	Feldberg and Smyth, 1977
grooming	0.1	rat	Gispen *et al*, 1976
prolactin release	0.17	rat	Dupont *et al*, 1977a
growth hormone release	1.7	rat	Dupont *et al*, 1977b
catalepsy	0.3 – 0.6	rat	Bloom *et al*, 1976; Jacquet and Marks, 1976

(1976) have proposed that tolerance to the opiates is due to enhancement of the intrinsic activity of adenylate cyclase on receptor stimulation; thus increasing amounts of inhibitory peptide are required to maintain a tolerable cAMP level. On removal of the receptor stimulus the basal cAMP level is above normal, giving rise to the withdrawal syndrome. C-Fragment has a high affinity (3nM when displacing H^3-naloxone in the presence of NaCl) for the opiate receptor (Bradbury *et al*, 1976b) which is due both to the N-terminal five residues (Tyr.Gly.Gly.Phe.Met-) and to the last four residues (...Lys.Lys.Gly.Gln); enkephalin produces 50% displacement of H^3-naloxone at a concentration of 90nM under the same conditions. It is remarkable that the enkephalin sequence, corresponding to the first five residues in a β turn conformation, shows a striking spatial resemblance to the alkaloid opiates oripavine and morphine (Bradbury *et al*, 1976c).

At the cellular level, microiontophoretic application of C-Fragment to neurons in the raphe magnus (Gent *et al*, 1977) and other brain areas (Guillemin, 1977) has produced inhibitory effects, except for the hippocampal pyramidal cells where excitatory effects have been observed. Microiontophoretic application of enkephalin produces inhibitory effects in the cortex, caudate and periaqueductal grey matter (Hill *et al*, 1976; Frederickson and Norris, 1976; Zieglgansberger *et al*, 1976) and in the brain stem (Gent and Wolstencroft, 1976), but excitatory effects on Renshaw cells in the spinal cord (Davies and Dray, 1976). Comparison of the potencies of enkephalin and C-Fragment in electrophysiological systems is difficult, but undoubtedly C-Fragment is significantly more potent than enkephalin in inhibiting the firing of single neurons (Gent *et al*, 1977). The inhibitory effects of the opiate peptide can also be seen when the turnover and release of other neurotransmitters are studied. In the *in vitro* assays on the guineapig ileum and mouse vas deferens the opiate peptides inhibit the electrically induced release of acetylcholine and noradrenaline respectively (Lord *et al*, 1977). Centrally, C-Fragment can also inhibit the turnover of acetylcholine (Moroni *et al*, 1977) and has been shown to inhibit the stimulated release of dopamine from striatal slices (Loh *et al*, 1976a) — in the hypothalamus this is thought to be the means by which C-Fragment stimulates secretion of prolactin, as the release of this hormone is under inhibitory dopamine control. In addition, in spinal trigeminal nuclei the opiate peptides have been reported to inhibit the stimulated release of Substance P (Jessell and Iversen, 1977).

The transient effects produced by the shorter peptides can be attributed, in part, to their short *in vivo* half life. The degradation of enkephalin is rapid, and is initiated by release of the N-terminal tyrosine by the action of endogenous aminopeptidases present in brain homogenates (Hambrook *et al*, 1976; Marks *et al*, 1977) and in brain membrane fractions (Meek *et al*, 1977; Miller *et al*, 1977; Austen and Smyth, 1977). However, even when enkephalin is protected from degradation its analgesic potency on intraventricular injection is less than 5% that of C-Fragment (Bradbury *et al*, 1977). The pentapeptide sequence has to be highly modified to produce analgesic potency approaching that of C-Fragment (Roemer *et al*, 1977). The high central potency of C-Fragment is due to a combination of its stability to degradation and its high receptor affinity induced by the last four residues in the sequence. The degradation of C-Fragment in striatal slices has been studied at concentrations approaching the physiological (Smyth and Snell, 1977), and has been shown to occur extracellularly to form γ and α endorphin and methionine enkephalin (Fig. 3). This suggests that the shorter peptides are degradation products of C-Fragment, but at the present time it is too early to speculate on the physiological relationship between C-Fragment and the shorter peptides.

Immunohistofluorescence methods applied to the opiate peptides are made difficult by the similarities in sequence of the different peptides. Specificity of an antiserum for a particular opiate peptide under immunoassay conditions cannot necessarily be extrapolated to

```
C-FRAGMENT      ————————>    γ-ENDORPHIN      ————————> α-ENDORPHIN
                                  │ ╲                  ╱  │
                                  │  ╲                ╱   │
                                  │   ╲              ╱    │
                                  │    ╲            ╱     │
                                  │     ╲          ╱      │
                                  │      ↘        ↙       │
                                  │   METHIONINE ENKEPHALIN    │
                                  │            │                │
                                  │            │                │
                                  │            ↓                │
                                  └————————> TYROSINE <————————┘
```

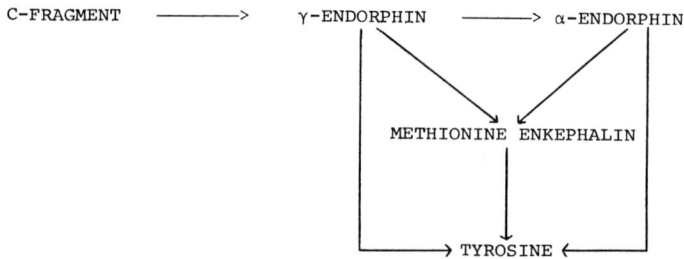

Fig. 3 Extracellular degradation pathways of C-Fragment.

specificity at the antibody concentrations required for immunohistofluorescence. As yet, this technique has been used only with antisera
to leucine and methionine enkephalin (Elde *et al*, 1976; Simatov *et al*,
1977), and has demonstrated immunoreactivity localised in neuronal
fibres and terminals corresponding closely to the distribution of opiate
receptors.

The analgesic and cataleptic effects of C-Fragment have potential
clinical relevance. Consequently, many pharmaceutical companies are
involved in the synthesis of enkephalin analogues, and no doubt, in
time, analogues will be devised that have advantages over morphine
in selectively producing analgesic, antitussive or antidiarrhoeal effects.
Promising results have already been reported for highly modified
enkephalin analogues with analgesic potency even after oral
administration (Roemer *et al*, 1977). It is unfortunate, however, that
the initial hope that the endogenous opiate peptides would not have
the addictive and tolerance problems of morphine has not been
realised: the endogenous peptides and synthetic analogues have been
shown to produce tolerance and/or cross tolerance with morphine
(Van Ree *et al*, 1976; Tsung *et al*, 1976; Wei and Loh, 1976; Hamper
et al, 1976; Waterfield *et al*, 1976; Szkely *et al*, 1977; Roemer *et al*,
1977). Whether the peptides have any advantageous differences from
morphine in producing fewer undesirable side effects awaits their
rigorous pharmacological screening. The catalepsy induced by
C-Fragment has been likened to that observed in some forms of
schizophrenia. Two alternative hypotheses have been advanced to
implicate disorders of C-Fragment levels in this disease. The first
postulates that schizophrenic symptoms are due to low levels of
C-Fragment, based on the observation that administration of
neuroleptics produces similar effects to C-Fragment and thus
alleviates the symptoms (Jacquet and Marks, 1976). The second
theory is that the catatonic symptoms characteristic of some forms

of schizophrenia are due to overproduction of C-Fragment (Bloom *et al*, 1976). Both these theories are testable by use of the opiate antagonist, naloxone, on schizophrenic patients, but as yet the results are equivocal (Terenius *et al*, 1976; Davis *et al*, 1977).

The opiate peptides are undoubtedly of fundamental importance as inhibitory modulators in the CNS, but an understanding of their physiological function must await further research.

Hypothalamic Hypophysiotropic Hormones

Thyroliberin (TRH), luliberin (LHRH) and somatostatin (GHIH) are the only hypothalamic hormones thus far to be isolated and chemically characterised. Until recently these peptides were thought to act exclusively on the anterior pituitary to control the release of hormones into the circulation. The development of specific antibodies to these peptides has made possible detailed analysis of their regional distribution in the CNS (Jackson and Reichlin, 1974; Oliver *et al*, 1974; Winokur and Utiger, 1974; Hokfelt *et al*, 1975c, d; Pelletier *et al*, 1974; Winokur *et al*, 1977; Kardon *et al*, 1977). As would be expected, such studies have shown that they are present in high concentration in the hypothalamus (up to 500 pmoles/g tissue). However, immunoreactivity is also present in nerve terminals in extrahypothalamic brain regions and the spinal cord, suggesting that these peptides might also be released from peptidergic neurons in the CNS. These peptides produce, in general, potent inhibitory effects on iontophoretic application to the cerebral and cerebellar cortex and the ventromedial nuclei of the hypothalamus (Dyer and Dyball, 1974; Renaud and Martin, 1975a, b; Wilber *et al*, 1976), although the basomedial hypothalamus has been reported to be excited by application of TRH (Dyball and Koizuma, 1969) and LHRH (Kawakami and Sakuma, 1974). The elegant work of Renaud and his colleagues (Wilber *et al*, 1976) has carefully mapped out the afferent and efferent pathways of the tuberoinfundibular secretory neurons of the hypothalamus. In addition to the neuro-secretory role at the median eminence, the neurons form collaterals synapsing within and outside the hypothalamus. Physiological states involving the secretion of the hypophysiotropic hormones from the median eminence may also involve release of the peptide from the collateral nerve terminals, producing a central effect in addition to the regulatory effect on the pituitary. Although these pathways could account for much of the extrahypothalamic levels of these peptides, the possibility that these peptides can also be biosynthesised extra-hypothalamically cannot be excluded. If these peptides are released from synapses they should act on specific receptors, but as yet TRH is the only hypothalamic hormone shown to have a high affinity central receptor (Burt and Snyder, 1975). At present, the physiological role of the hypothalamic hormones in controlling release of pituitary hormones is considerably better understood than their probably equally important physiological function within the CNS.

Conclusion

The peptides listed in Table I are distributed unevenly in neural tissue, and may be considered candidates for neurotransmitter or neuro-secretory roles. Establishment will follow the fulfillment of the criteria required for neurotransmitters, and will come from un-ambiguous isolation, and study of their regional distribution, central effects and degradation. Thus far, however, only Substance P and the opiate peptides have been subject to sufficient study to consider them acceptable candidates. The peptides differ from the low molecular weight neurotransmitters in being biosynthesised ribosomally and in not being subject to uptake and recycle processes at the synapse. Nevertheless, the peptides appear to serve inhibitory or excitatory roles similar to those observed for the low molecular weight trans-mitters. Each peptide fulfils its role specified by the unique com-plementary of its amino acid sequence with the physiological receptor. The response produced by the sequence of events following receptor binding will depend on the location and type of neuron on which the receptor is situated. Undoubtedly peptidergic neurons represent a new dimension to neurophysiology and neuropharmacology and are one of the most promising and exciting areas for academic and clinician alike.

References

Austen, B.M. and Smyth, D.G. (1977). *Biochem. Biophys. Res. Commun.* **76**, 477.

Bajusz, S., Ronai, A.Z., Szekely, J.I., Graf, L., Dunai-Kovacs, Z. and Berzetei, I. (1977). *Febs. Letts.* **76**, 91.

Belluzzi, J.D., Grant, N., Garsky, V., Sarantakis, D., Wise, C.D. and Stein, L. (1976). *Nature* **260**, 625.

Bloom, F., Segal, D., Ling, N. and Guillemin, R. (1976). *Science* **194**, 630.

Bradbury, A.F., Smyth, D.G. and Snell, C.R. (1975). *In* "Peptides: Chemistry Structure and Biology" (ed. Meineinhofer, J.), p. 609. Ann Arbor Sci.

Bradbury, A.F., Feldberg, W.S., Smyth, D.G. and Snell, C.R. (1976a). *In* "Opiates and endogenous opioid peptides" (ed. Kosterlitz, H.W.), p. 9. Elsevier.

Bradbury, A.F., Smyth, D.G., Snell, C.R., Birdsall, N.J.M. and Hulme, E.C. (1976b). *Nature* **260**, 793.

Bradbury, A.F., Smyth, D.G. and Snell, C.R. (1976c). *Nature* **260**, 165.

Bradbury, A.F., Smyth, D.G., Snell, C.R., Deakin, J.F.W. and Wendlandt, S. (1977). *Biochem. Biophys. Res. Commun.* **74**, 748.

Brownstein, M.J., Mroz, E.A., Kizer, J.S., Palkovits, M. and Leeman, S.E. (1976). *Brain Research* **116**, 299.

Burt, D.R. and Snyder, S.H. (1975). *Brain Research* **93**, 309.

Buscher, H.H., Hill, R.C., Romer, D., Cardinaux, F., Closse, A., Hauser, D. and Pless, J. (1976). *Nature* **261**, 423.

Chang, M.M. and Leeman, S.E. (1970). *J. Biol. Chem.* **245**, 4784.

Chang, M.M., Leeman, S.E. and Niall, H.D. (1971). *Nature New Biol.* **232**, 86.

Collier, H.O.J. and Roy. (1977). *Biochem. Soc. Trans.* (in press).

Coy, D.H., Kastin, A.J., Schally, A.V., Morin, O., Caron, N.G., Labrie, F., Walker, J.M., Fertel, R., Bernston, G.G. and Sandman, C.A. (1976). *Biochem. Biophys. Res. Comm.* **73**, 632.

Cusan, L., Dupont, A., Kledzik, G.S., Labrie, F., Coy, D.H. and Schally, A.V. (1977). *Nature* **268**, 544.

Davies, J. and Dray, A. (1976). *Nature* **262**, 603.

Davis, G.C., Bunney, W.E., Defraites, E.G., Kleinman, J.E., van Kammen, D.P., Post, R.M. and Wyatt, R.J. (1977). *Science* **197**, 74.

Dupont, A., Cusan, L., Labrie, F., Cox, D.H. and Li, C.H. (1977a). *Biochem. Biophys. Res. Commun.* **75**, 76.

Dupont, A., Cusan, L., Garon, M., Labrie, F. and Li, C.H. *Proc. Natl. Acad. Sci. U.S.A.* **74**, 388.

Dyball, R.E. and Koizuma, K. (1969). *J. Physiol. (London)* **201**, 711.

Dyer, R.G. and Dyball, R.E.J. (1974). *Nature* **252**, 486.

Elde, R., Hökfelt, T., Johansson, O. and Terenius, L. (1976). *Neuroscience* **1**, 349.

Emson, P.C., Kanazawa, I., Cuello, A.C. and Jessell, T.M. (1977). *Biochem. Soc. Trans.* **5**, 187.

Feldberg, W.S. and Smyth, D.G. (1976). *J. Physiol.* **260**, 30p.

Feldberg, W.S. and Smyth, D.G. (1977). *Brit. J. Pharm.* **60**, 445.

Frederickson, R.C.A. and Norris, F.H. (1976). *Science* **194**, 440.

Gent, J.P., Smyth, D.G., Snell, C.R. and Wolstencroft, H. (1977). *Brit. J. Pharm.* (in press).

Gent, J.P. and Wolstencroft, J.H. (1976). *Nature* **261**, 426.

Gispen, W.H., Wiegant, V.M., Bradbury, A.F., Hulme, E.C., Smyth, D.G., Snell, C.R. and de Wied, D. (1976). *Nature* **264**, 794.

Graf, L., Barat, E. and Patthy, A. (1976). *Acta. Biochim. biophys. Acta. Sci., (Hung.)* **11**, 121.

Graf, L., Szekely, J.I. Ronai, A.Z., Dunai-Kovacs, S.Z. and Bajusz, S. (1976). *Nature* **263**, 240.

Guillemin, R. (1977). *In* "Proceedings of the VI International Congress of Endocrinology". Elsevier, North Holland (in press).

Hambrook, J.M., Morgan, B.A., Rance, M.J. and Smith, C.F.C. (1976). *Nature* **262**, 782.

Henry, J.L. (1976). *Brain Research* 114, 439.
Hill, R.G., Pepper, C.M. and Mitchell, J.F. (1976). *Nature* 262, 604.
Hökfelt, T., Elde, R., Johansson, O., Luft, R. and Arimura, A. (1975b). *Neuroscience Letts.* 1, 231.
Hökfelt, T., Elde, R., Johansson, O., Luft, R., Nilsson, G. and Arimura, A. (1976). *Neuroscience* 1, 131.
Hökfelt, T., Fuxe, K., Johansson, O., Jeffcote, S. and White, N. (1975c). *Neuroscience Letts.* 1, 133.
Hökfelt, T., Fuxe, K., Johansson, O., Jeffcote, S. and White, N. (1975d). *Eur. J. Pharmacol.* 34, 389.
Hökfelt, T., Kellerth, J-O., Nilsson, G. and Pernow, B. (1975a). *Brain Research* 100, 235.
Hughes, J., Smith, T.W., Kosterlitz, H.W., Fothergill, L.A., Morgan, B.A. and Morris, H.R. (1975). *Nature* 258, 577.
Iversen, L.L., Jessell, T. and Kawazawa, T. (1976). *Nature* 264, 81.
Jackson, I.M.D. and Reichlin, S. (1974). *Endocrinology* 75, 854.
Jacquet, Y.F. and Marks, N. (1976). *Science* 194, 634.
James, T. and Starr, M.S. (1977). *J. Pharm. Pharmac.* 29, 181.
Jessell, T.M. and Iversen, L.L. (1977). *Nature* 268, 549.
Kanazawa, I. and Jessell, T. (1976). *Brain Research* 117, 362.
Kardon, F.C., Winokur, A. and Utiger, R.D. (1977). *Brain Research* 122, 578.
Kawakami, M. and Sakuma, Y. (1974). *Neuroendocrinology* 15, 290.
Konishi, S. and Otsuka, M. (1974). *Brain Research* 65, 397.
Krieger, D.T., Liotta, A., Suda, T., Palkovits, M. and Brownstein, M.J. (1977). *Biochem. Biophys. Res. Commun.* 76, 930.
Lampert, A., Nirenberg, M. and Klee, W.A. (1976). *Proc. Natl. Acad. Sci. U.S.A.* 73, 3165.
Leeman, S.E. and Hammerschlag, R. (1967). *Endocrinology* 81, 803.
Lembeck, F. (1953). *Naun-Schmeid. Arch. Exp. Path. Pharm.* 219, 197.
Lembeck, F. and Starke, K. (1968). *Naun-Schmeid. Arch. Exp. Path.* 259, 375.
Li, C.H. and Chung, D. (1976). *Proc. Natl. Acad. Sci. U.S.A.* 73, 1145.
Loh, H.H., Brase, D.A., Sampath-Kharma, S., Mar. J.B., Way, E.L. and Li, C.H. (1976a). *Nature* 264, 567.
Loh, H.H., Tseng, L.F., Wei, E. and Li, C.H. (1976b). *Proc. Natl. Acad. Sci. U.S.A.* 73, 2895.
Lord, J.A.H., Waterfield, A.A., Hughes, J. and Kosterlitz, H.W. (1977). *Nature* 267, 495.
Marks, N., Grynbaum, A. and Neidle, A. (1977). *Biochem. Biophys. Res. Comm.* 74. 1552.
Martin, J.B., Renaud, L.P. and Brazeau, P. (1975). *Lancet* ii, 393.
Meek, J.L., Yang, H-Y.T. and Costa, E. (1977). *Neuropharmacology* 16, 151.
Miller, R.J., Chang, K-J., Cuatrecasas, P. and Wilkinson, S. (1977). *Biochem. Biophys. Res. Commun.* 74, 1311.
Moroni, F., Cheney, D.L. and Costa, E. (1977). *Nature* 267, 267.
Mroz, E.A., Brownstein, M.J. and Leeman, S.E. (1976). *Brain Research* 113, 597.

Mroz, E.A., Brownstein, M.J. and Leeman, S.E. (1977). *Brain Research* **125**, 305.

Nicoll, R.A. (1976). *Neuroscience Symp.* **1**, 99.

Nilsson, G., Hökfelt, T. and Pernow, B. (1974). *Med. Biol.* **52**, 424.

Oliver, C., Eskay, R.L., Ben-Jonathan, N. and Porter, J.C. (1974). *Endocrinology* **96**, 540.

Otsuka, M., Konishi, S. and Takahashi, T. (1975). *Fed. Proc.* **34**, 1922.

Otsuka, M. and Konishi, S. (1976). *Nature* **264**, 83.

Olpe, H-R. and Koella, W.P. (1977). *Brain Research* **126**, 576.

Palay, V-C. and Palay, S.L. (1977). *Proc. Natl. Sci. U.S.A.* **74**, 3597.

Pelletier, G., Labrie, F., Puviani, R., Avimura, A. and Schally, A.V. (1974). *Endocrinology* **95**, 314.

Pert, C.B., Pert, A., Chang, G.J-K. and Fong, B.T.W. (1976). *Science* **194**, 330.

Pickel, V.M., Reis, D.J. and Leeman, S.E. (1977). *Brain Res.* **122**, 534.

Powell, D., Leeman, S.E., Tregear, G.W., Niall, H.D. and Potts, J.T. Jnr. (1973). *Nature New Biol.* **241**, 252.

Renaud, L.P. and Martin, J.B. (1975a). *Nature* **255**, 233.

Renaud, L.P. and Martin, J.B. (1975b). *Brain Research* **86**, 150.

Roemer, D., Beuscher, H.H., Hill, R.C., Pless, J., Bauer, W., Cardinaux, F., Closse, A., Hauser, D. and Heguenin, R. (1977). *Nature* **268**, 547.

Schenker, C., Mroz, E.A. and Leeman, S.E. (1976). *Nature* **264**, 790.

Simantov, R., Kuhar, M.J., Uhl, G.R. and Snyder, S.H. (1977). *Proc. Natl. Acad. Sci. U.S.A.* **74**, 2167.

Simantov, R. and Snyder, S.H. (1976). *Proc. Natl. Acad. Sci. U.S.A.* **73**, 2515.

Smyth, D.G. and Snell, C.R. (1977). *Febs. Letts.* **78**, 225.

Szkely, J.I., Ronai, A.Z., Dunai-Kovacs, Z., Miglecz, E., Bajusz, S. and Graf, L. (1977). *Life Sci.* **20**, 1259.

Terenius, L., Wahlstrom, A., Lindstrom, L. and Widerlov, E. (1976). *Neuroscience Letts.* **3**, 157.

Tsung, L-F., Loh, H.H. and Li, C.H. (1976). *Proc. Natl. Acad. Sci. U.S.A.* **73**, 4187.

von Euler, U.S. and Gaddum, J.H. (1931). *J. Physiol.* **72**, 74.

van Ree, J.M. de Wied, D., Bradbury, A.F., Hulme, E.C., Smyth, D.G. and Snell, C.R. (1977). *Nature* **264**, 792.

Waterfield, A.A., Hughes, J. and Kosterlitz, H.W. (1976). *Nature* **260**, 624.

Wei, E. and Loh, H. (1976). *Science* **193**, 1262.

Wilber, J.F., Montoya, E., Plotnikoff, N.P., White, W.F., Genrich, R., Renaud, L.P. and Martin, J.B. (1976). *Rec. Prog. Hormone Res.* **32**, 117.

Winokur, A., Davis, R. and Utiger, R.D. (1977). *Brain Research* **120**, 423.

Winokur, A. and Utiger, R.D. (1974). *Science* **185**, 265.

Zetler, G. (1976). *Biochem. Pharm.* **25**, 1817.

Zieglgänsberger, W., Fry, J.P., Herz, A., Moroder, L. and Wunsch, E. (1976). *Brain Research* **115**, 160.

SUBJECT INDEX